T5-AWI-654

Advances in the Stabilization and Controlled Degradation of Polymers

International Conference on

Advances in the Stabilization and Controlled Degradation of Polymers

Volume II

Edited by

Angelos V. Patsis

Head, Chemistry Department
State University of New York
New Paltz, New York

Presented in Lucerne, Switzerland
May 1987

Lancaster · Basel

Published in the Western Hemisphere by
Technomic Publishing Company, Inc.
851 New Holland Avenue
Box 3535
Lancaster, Pennsylvania 17604 U.S.A.

Distributed in the Rest of the World by
Technomic Publishing AG

Printed in the United States of America
10 9 8 7 6 5 4 3 2 1

Main entry under title:
International Conference on Advances in the Stabilization and Controlled Degradation of Polymers—Volume II

A Technomic Publishing Company book
Bibliography: p.

ISSN No. 1042-3982
ISBN No. 87762-588-3

TABLE OF CONTENTS

D. R. Bauer[1]

Photodegradation Chemistry in Thermoset Coatings and Stabilization by Hindered Amines

ABSTRACT

Photodegradation chemistries in acrylic copolymers crosslinked with melamine formaldehyde resins and isocyanate resins are compared. Particular emphasis is given to the interactions between free radical oxidation and crosslink degradation. In the case of coatings crosslinked with melamine formaldehyde resins, formaldehyde related chemistry is found to be critical. The effects of the addition of hindered amine light stabilizers on the photooxidation rates are discussed. Stabilization chemistry is followed through measurements of hindered amine and nitroxide concentrations as a function of exposure time, exposure composition, and coating composition. A model for the photostabilization chemistry of hindered amines in crosslinked coatings is presented, which attempts to account for the differences in effectiveness and permanence observed for hindered amines in melamine and urethane coatings.

KEY WORDS

Photodegradation, photostabilization, coatings, urethanes, melamine formaldehyde resins, acrylics, crosslink structure, hindered amine light stabilizer.

INTRODUCTION

Hindered amine light stabilizers are widely used to suppress photodegradation in polymers used outdoors. Most of the studies of hindered amine stabilizer mechanisms have focused on model compounds or thermoplastic polymers such as polypropylene [1–6]. Oxidation in unstabilized polypropylene is characterized by a relatively short induction period followed by rapid oxidation. The kinetic chain length can be quite long due to efficient propagation reactions down the polymer chain. The addition of hindered amines brings dramatic results. In typical accelerated ultraviolet (UV) exposures, the induction period increases from 50 to 2000 hours [6]. The basics of hindered amine stabilization is thought to involve oxidation of amine groups to nitroxides, reaction of nitroxides with radicals to form aminoethers and recycling of the aminoether back to nitroxide [7]. These reactions suppress photodegradation by interfering with free radical propagation. To account for the observed effectiveness of hindered amines in polypropylene, other factors including association of hindered amines with hydroperoxides [8,9] and excited state quenching have also been considered [10].

Although photodegradation chemistry in crosslinked coatings is also based on free radical photooxidation, the specific chemistries and kinetics are significantly different from those in polypropylene. Instead of having an induction period followed by rapid oxidation, photooxidation in coatings appears to be relatively constant in time [11]. The kinetic chain lengths are relatively short (< 10). There are specific degradation chemistries which affect the crosslinked network structure, causing changes in the physical properties of the coating [12–16]. In view of the differences in degradation chemistry, it is not surprising that the addition of hindered amine to coatings will result in different stabilization effects than are observed in polypropylene. This paper summarizes differences observed in the effectiveness and permanence of one hindered amine (bis-(2,2,6,6-tetramethyl-4-piperidinyl) sebacate) as a function of coating composition and exposure variables [11,13,16]. Differences in the rate of consumption of hindered amine and in the nitroxide concentration are also discussed [17–19]. An attempt is made to account for the observed behavior on the basis of specific differences in degradation and stabilization reactions.

EXPERIMENTAL

Coating Materials

The coatings studied in this work consist of hydroxy functional acrylic copolymers crosslinked with a melamine formaldehyde resin or an isocyanate resin. The acrylic copolymers were prepared by conventional free

[1]Research Staff, Ford Motor Company, P.O. Box 2053, Dearborn, MI 48121.

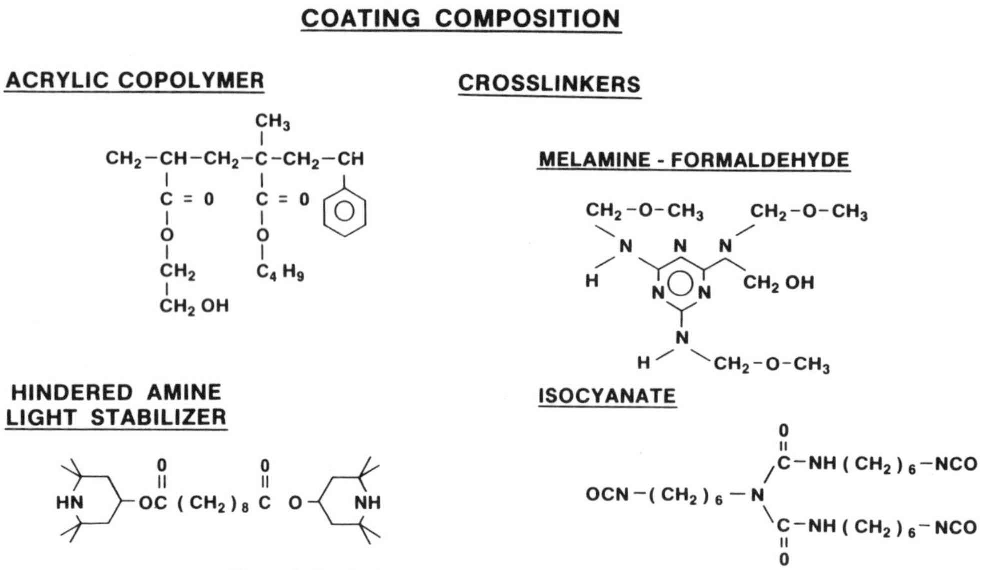

COATING COMPOSITION

ACRYLIC COPOLYMER

CROSSLINKERS

MELAMINE - FORMALDEHYDE

HINDERED AMINE LIGHT STABILIZER

ISOCYANATE

Figure 1. Typical components of thermoset coatings studied.

radical copolymerization and had molecular weights ranging from 1500–8000 [20]. Most of the studies used polymers whose molecular weight was suitable for use as high solids automotive topcoats (i.e., $M_n \sim 2000$). The melamine formaldehyde resin used was a partially alkylated melamine (Cymel 325 from American Cyanamid). Coatings were formulated using a ratio of polymer to crosslinker of 70:30 and cured for 20 minutes at 130°C. The isocyanate crosslinker was a biuret of hexamethylene diisocyanate (L2291A from Mobay). The urethane coatings were formulated to obtain a 1:1 hydroxy to isocyanate ratio. No external catalysts were used. Coatings were cured for 20 minutes at 130°C. The hindered amine light stabilizer (HALS-I) was obtained from Ciba-Geiby (TIN-770) and recrystallized before use. Typical structures of the polymers, crosslinkers, and hindered amine are shown in Figure 1.

Exposure Conditions

Samples were exposed in modified Atlas UV-2 exposure chambers. The light source consisted of two FS-20 UV-A fluorescent bulbs. Light intensity was varied by the use of neutral density filters. The chambers were modified to allow independent control of air temperature and dew point [20]. The air temperature was maintained at 60°C while dew points ranged from −40°C to 50°C. A dark condensing humidity cycle was generally not employed.

Experimental Methods

Photodegradation chemistry in these coatings has been studied by infrared spectroscopy and magic angle nuclear magnetic resonance. The details of these techniques are described elsewhere [12–16]. Degradation rates were determined by following changes in band intensities with exposure time. For example, the rate of melamine crosslink scission was followed by measuring the rate of disappearance of the melamine methoxy band at 915 cm^{-1} in the infrared. Comparisons of these rates with and without hindered amine were used to determine hindered amine effectiveness. The concentration of hindered amine was determined as a function of exposure time by extracting the hindered amine from the coating and measuring the concentration in the extract by gas chromatography [18]. The concentration of nitroxide was determined using electron spin resonance (ESR) [17,21]. Quantification details have been described elsewhere [22]. In some cases, another radical species was also observed in ESR spectra of the coating. This radical had a much different lineshape than the nitroxide and both could be quantified. The nature of the other radical component(s) is unknown though the signal is most likely due primarily to peroxy species. ESR was also used to measure the photoinitiation rates of free radicals in these coatings [23,24]. Instead of doping with a hindered amine, the coatings were doped with a persistent nitroxide and the photoinitiation rate determined from the rate of disappearance of nitroxide during photolysis. This technique will be referred to as the nitroxide decay assay to differentiate it from the HALS doping studies.

RESULTS AND DISCUSSION

Basic Degradation and Stabilization Kinetics

As will be discussed in more detail below, photodegradation chemistry in these coatings is dominated by free radical oxidation. Free radical oxidation consists of three steps: initiation, propagation and termination [25]. Hindered amines function by competing with the propagation step and thus shortening the oxidation chain length. A general scheme for the degradation and stabilization reactions are presented below:

Initiation

$$A + h\nu \xrightarrow{W_i} Y\cdot \quad (1)$$

Propagation

$$Y\cdot + O_2 \xrightarrow{k_{OX}} YOO\cdot \quad (2)$$

$$YOO\cdot + YH \xrightarrow{k_{YH}} Y\cdot + YOOH \quad (3)$$

Termination

$$2\ YOO\cdot \xrightarrow{k_T} \text{products} \quad (4)$$

Chain Branching

$$YOOH + h\nu \xrightarrow{k_{CB}} YO\cdot + \cdot OH \quad (5)$$

Hydroperoxide Decomposition

$$YOOH \xrightarrow{k_{NCB}} \text{products} \quad (6)$$

Nitroxide Formation

$$NH + YOO\cdot \xrightarrow{k_{NH}} NO\cdot \quad (7)$$

Nitroxide Scavenging

$$NO\cdot + Y\cdot \xrightarrow{k_{NO}} NOY \quad (8)$$

Nitroxide Recycling

$$NOY + YOO\cdot \xrightarrow{k_{NOY}} NO\cdot \quad (9)$$

Nitroxide Excited State Chemistry

$$NO\cdot + h\nu\ (\text{or } A^*) \xrightarrow{k_{EX}} NO^* \quad (10)$$

$$NO^* + YH \xrightarrow{k_{ABS}} NOH + Y\cdot \quad (11)$$

where A is a chromophore in the coating which absorbs light to form free radicals, Y·, at a rate W_i and NH, NO·, and NOY represent amine, nitroxide, and aminoether functionality respectively which is associated with the hindered amine. Decomposition of hydroperoxides has been discussed by Ingold [26]. Excited state nitroxides can be formed either by direct absorption [27] or by energy transfer from another excited state species [28]. Excited state nitroxides are effective hydrogen atom abstractors [27,29].

The effectiveness of a hindered light stabilizer depends on the concentration of the different stabilizing species and on the following ratios of rate constants: k_{NH} and k_{NOY} to k_{YH} and k_{NO} to k_{OX}. The oxidation chain length is also important. In general, the longer the oxidation chain length the more likely the hindered amine will be effective. This, in part, explains why hindered amines are so effective in polypropylene. In order to derive values for the stabilization rate constants it is necessary to determine the rate constants for Reactions (1)–(7). The photoinitiation rate, W_i, can be measured directly using the nitroxide decay assay described above. This assay also provides a direct measure of the importance of nitroxide excited state chemistry [Reactions (10) and (11)]. By measuring the nitroxide decay kinetics at low nitroxide concentration, it is possible to determine the quantity k_{NO}/k_{OX} and the importance of chain branching. Combining these data with infrared spectroscopic estimates of the amount of oxidation and ESR measurements of the peroxy radical concentration yield values for k_{YH} and k_T. Typical values have been given elsewhere [19]. Chain branching is found to be unimportant since the oxidation chain lengths are relatively short and there are apparently effective means to decompose hydroperoxides in the coatings. The value of k_{NO}/k_{OX} in these coatings is roughly 0.3. This value is consistent with solution studies of the relative reactivity of nitroxide and oxygen with radicals [1]. For the exposure conditions used here, oxidation chain lengths typically ranged from 3–10. The rate constants can be expected to vary from coating to coating and may even vary with exposure conditions if the exposure conditions affect the nature of the free radical Y·.

Coating Photodegradation Chemistry

Typical infrared spectra of undegraded and degraded acrylic/melamine and acrylic/urethane coatings are shown in Figures 2 and 3, respectively [14,16]. In both coatings there is a broadening and increase in intensity (relative to the hydrocarbon band at 2950 cm^{-1}) in the carbonyl band at 1730 cm^{-1}. This broadening is indicative of the formation of oxidation products. Difference spectra reveal a band at ~1700 cm^{-1}, which is likely due to formation of carboxylic acids. It is difficult to identify specific oxidation products in either the infrared or NMR spectra. The increase in area of the carbonyl band can be measured as a function of exposure time. The area increases roughly linearly with time [11]. The rate of increase is used to estimate oxidation chain lengths. The apparent lack of autocatalytic oxidation is consistent with the lack of hydroperoxide chain branching indicated by the nitroxide decay assay [19]. Thus, coating oxidation

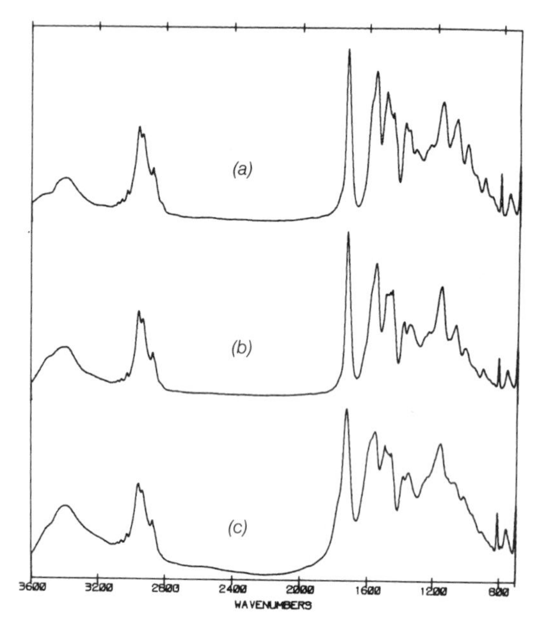

Figure 2. Infrared spectra of melamine formaldehyde crosslinked coatings. Spectra taken from unexposed sample (a); sample exposed to condensing humidity (b); and sample exposed to UV light and condensing humidity (c). Figure reprinted from [14].

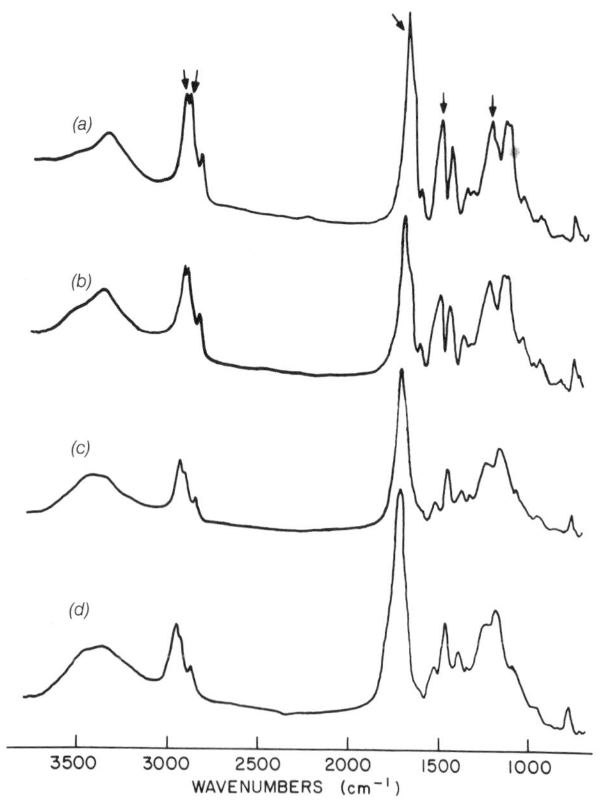

Figure 3. Infrared spectra of urethane crosslinked coatings. Spectra taken from unexposed sample (a); sample exposed to condensing humidity (b); sample exposed to UV light and condensing humidity (c); and sample exposed to UV light at a dew point of $-40°C$ (d). Figure reprinted from [16].

kinetics are in sharp contrast to the oxidation kinetics in polypropylene. The rate of oxidation in an unstabilized melamine coating is similar ($\pm 50\%$) to that in an unstabilized urethane coating if the same acrylic polymer is used. The oxidation rate in urethane coatings is found to be independent of humidity while the rate in melamine coatings increases with increasing humidity (almost doubling over the dew point range -40 to $50°C$) [11,13]. From measurements of the photoinitiation rate of free radicals in these coatings, it can be concluded that the copolymer largely determines the initiation rate and that the initiation rate is independent of humidity. This suggests that the increase in oxidation observed with increasing humidity during UV exposure in melamine coatings is due to an increase in the propagation rate. A possible explanation for this effect is suggested below.

In addition to general oxidation, chemical changes which are specific to the crosslinked structure are also observed. In the urethane coating, for example, there is clear loss of the amide II and amide IV bands (at 1540 and 1240 cm^{-1}) [16]. This suggests scission of the acrylic-urethane crosslink. The scission rate is independent of humidity and thus is not a hydrolytic reaction. Scission appears to be caused by free radical attack on the urethane crosslink (possibly H-atom abstraction of one of the urethane hydrogens) [30]. For some acrylic polymers, after the urethane links are broken, the crosslinker is lost from the coating. The extent of loss of crosslinker increases somewhat with increasing exposure humidity even though the rate of scission does not.

Acrylic-melamine crosslinks are also broken during exposure to UV light and humidity. For these crosslinks, the rate of scission increases with both increasing UV light intensity and humidity during the exposure [13]. In the absence of UV light, the crosslinks undergo a slow hydrolysis [12]. The mechanism for this acid catalyzed hydrolysis has been studied [31]. When acrylic-melamine crosslinks are hydrolyzed, melamine methylol groups are formed. These methylol groups can either deformylate to amine or self-condense to form a melamine-melamine crosslink (appearance of a band at 1360 cm^{-1}) also releasing formaldehyde. Emission of formaldehyde has been observed using gas phase infrared spectroscopy [32]. Acrylic-melamine crosslinks are also broken on exposure to UV light in the absence of humidity [13]. The rate depends on light intensity and the mechanism appears to involve free radical attack on the crosslink. No melamine-melamine crosslink formation is observed suggesting that no melamine methylol groups are formed. On exposure to both UV light and humidity, the rate of crosslink scission increases to a value greater than the sum of the rates for UV light only and humidity only exposures [13]. Melamine-melamine crosslink formation is observed [14,15]. The rate of formaldehyde release is also greater in this exposure than for the UV light only or humidity only exposures [32]. A reaction scheme which

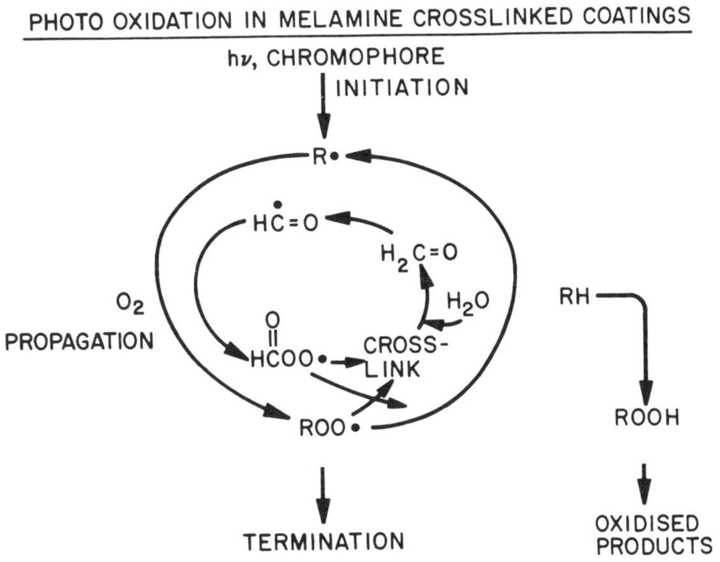

Figure 4. Degradation chemistry of melamine crosslinked coatings.

accounts for these observations has been proposed [32]. According to this scheme, free radical attack on the crosslink forms a species which when reacted with water produces a melamine methylol group. The methylol group then can self-condense to form a melamine-melamine crosslink and release formaldehyde into the coating. Comparison of the rate of scission and the rate of formaldehyde release suggests that most of the formaldehyde formed remains in the coating. Retained formaldehyde can participate in the free radical oxidation kinetics via formation of performic acid. Peracids are strong oxidants and could account for the increase in oxidation with humidity observed in the melamine coating. A schematic picture of the interaction of humidity, formaldehyde, and UV light in the degradation of melamine crosslinked coatings is shown in Figure 4.

Effectiveness of Hindered Amines

Experimentally, the effectiveness of a hindered amine light stabilizer can be measured by comparing the degradation rates with and without hindered amine. In typical high solids acrylic/melamine coatings, the addition of 2% by weight hindered amine reduces the oxidation and acrylic-melamine crosslink scission rate by a factor of about 2 [11,13]. In acrylic/urethane coatings, the effects are much more dramatic [30]. The oxidation rate can be reduced by factors of 5–10 while the acrylic-urethane crosslink scission rate can be reduced by factors of 10–20, Figure 5. The rate of oxidation in acrylic/urethane coatings appears to be reduced almost to the level of the photoinitiation rate (i.e., an oxidation chain length of 1). The effectiveness of hindered amines in these coatings appears to be roughly constant for long periods of exposure time. This constant effectiveness and the big difference in effectiveness between urethane and melamine coatings is not easy to explain. In the sections that follow, studies of HALS chemistry will be presented which are aimed at understanding the chemistry underlying these observations and at interpreting these observations in terms of changes in the different stabilization rate constants. Formaldehyde chemistry will play a key role in these discussions.

Rate of HALS Consumption

According to Reaction (7), hindered amine functionality is consumed at a rate determined by k_{NH} and the

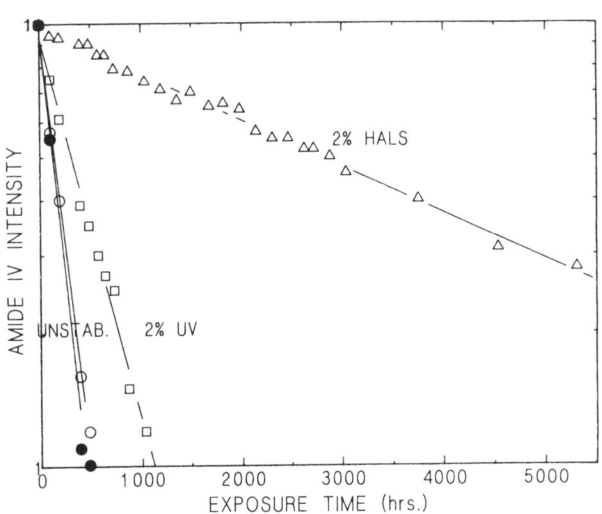

Figure 5. Effect of HALS I on rate of loss of Amide II signal in an acrylic/urethane coating. Figure reprinted from [30].

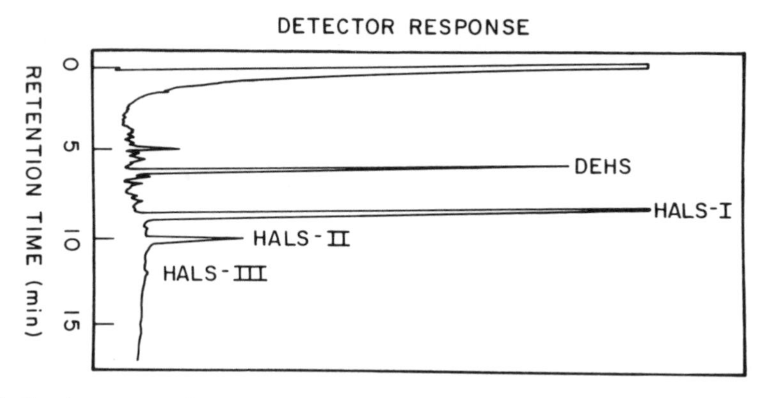

Figure 6. Gas chromatogram from extracts of cured acrylic/melamine coating. Figure reprinted from [18].

steady state concentration of YOO·. The peroxy radical concentration can be measured using ESR. Thus, in principle k_{NH} can be determined from measurements of the loss of amine functionality. A gas chromatography technique to determine the amine concentration has been described and a typical gas chromatogram of an extract from a cured acrylic/melamine coating is shown in Figure 6 [18]. In addition to the signal from HALS I, two weaker signals are observed which can be assigned to the mono- and di-N-methylated versions of HALS I. Approximately 15% of the NH groups are converted to $N-CH_3$ groups during cure of the melamine coating. This reaction does not occur in the urethane coating. This observation can be explained by the fact that formaldehyde is released during the cure of melamine coatings. Formaldehyde is a known methylating agent for secondary amines [33]. There is more than enough formaldehyde released during melamine crosslinking to account for the observed methylation.

The concentration of HALS I (and the other hindered amines) has been measured as a function of exposure time for both a melamine and urethane crosslinked acrylic copolymer doped with 2% by weight HALS I at different light intensities and humidities [18]. The photoinitiation rate of free radicals for both coatings at the maximum light intensity used was 12×10^{-8} mole/g·min. The oxidation chain length was ~3–5. At this intensity, the initial rate of consumption of NH functionality was 0.78×10^{-8} mole/g·min for the urethane coating independent of humidity. The initial rate for melamine coating was 1.35×10^{-8} mole/g·min at a dew point of $-40°C$ and 0.59×10^{-8} mole/g·min at a dew point of 25°C. In the urethane coating NH functionality is lost steadily with almost 75% consumed by 140 hours of exposure. In the melamine coating under dry exposure the consumption of NH and $N-CH_3$ groups is even more rapid with roughly 90% of the NH groups consumed by 140 hours. The rate of loss of NH functionality is roughly proportional to the square root of the light intensity. The behavior in the melamine coating is quite different under humid exposures. The concentration of NH groups drops 40% in the first 40 hours then remains roughly constant out to over 300 hours. The $N-CH_3$ concentration actually increases initially then parallels the NH concentration. The fact that the NH concentration reaches a plateau value suggests that some mechanism exists to recycle nitroxide or aminoethers back to NH groups. The observed behavior can be accounted for by the following reaction scheme:

Urethane Coating

$$\text{YOO}\cdot + \text{NH} \rightarrow \text{NO}\cdot \quad k_{NH'} \tag{12}$$

Melamine Coating

$$H_2C{=}O + \text{YOO}\cdot \rightarrow \text{HC}\cdot({=}O) \tag{13}$$

$$\text{HC}\cdot({=}O) + O_2 \rightarrow \text{HC}({=}O)\text{OO}\cdot \tag{14}$$

$$\text{YOO}\cdot + \text{NH} \rightarrow \text{NO}\cdot \quad k_{NH'} \tag{15}$$

$$\text{HC}({=}O)\text{OO}\cdot + \text{NH} \rightarrow \text{NO}\cdot \quad k_{NH''} \tag{16}$$

$$\text{NO}\cdot + \text{HC}\cdot({=}O) \rightarrow \text{NOC}({=}O)\text{H} \tag{17}$$

$$\text{NOC}({=}O)\text{H} \rightarrow \text{NH} + CO_2 \quad k_{RE} \tag{18}$$

(note: water is required as a proton transfer agent, see [18])

$$\text{NH} + H_2C{=}O \rightarrow N{-}CH_3 \tag{19}$$

In the urethane case, the initial rate of NH consumption is $k_{NH'}$ [NH] [YOO·]. The measured rate k_{NH} is equal to $k_{NH'}$. In the case of the melamine under dry conditions the rate is $k_{NH'}$ [NH] [YOO·] + $k_{NH''}$ [NH] [HC(=O)OO·]. The measured rate constant k_{NH} in the melamine coating will be a weighted average of $k_{NH'}$ and $k_{NH''}$. The total radical concentration is similar in the melamine and urethane coating. Peracids are much stronger oxidants of NH groups than are peroxy radicals [3]. Thus, $k_{NH''} > k_{NH'}$ and the rate of consumption of

NH groups is greater in the melamine coating under dry conditions than it is in the urethane coating. Under humid conditions, however, the recycling reaction [Reaction (18)] occurs and the consumption rate of amine is reduced by the quantity k_{RE} [NOC(=O)H]. Thus, the rate of consumption of NH groups can decrease even though the amount of formaldehyde released increases. When recycling occurs, it is not possible to determine k_{NH} from the amine consumption data. A determination of k_{NH} using the initial rate of formation of nitroxide will be given below. Formaldehyde released in the melamine coating on exposure also accounts for the high level of $N-CH_3$ functionality maintained in this coating [Reaction (19)].

Although clear differences in the conversion of hindered amine to nitroxide can be observed in the different coatings, the overall effect of these differences on the effectiveness of the stabilizer may be small. $N-CH_3$ groups and NH groups are converted to nitroxide at a similar rate. In all cases the conversion of NH groups to nitroxide is inefficient relative to the photoinitiation rate or oxidation rate (i.e., the ratio k_{NH} [NH]/k_{YH} [YH] is small). Thus, Reaction (7) can not be responsible for the observed reduction in oxidation chain length observed in these coatings.

Nitroxide Concentration Studies

Since Reaction (7) is not the main stabilizing reaction, it is necessary to study Reactions (8) and (9) which involve nitroxides and aminoethers. It is not possible to quantify the aminoether compositions and concentration in crosslinked coatings. The nitroxide concentration, on the other hand, can be quantified [21,22]. As shown in Figure 7, the concentration of nitroxide in hindered amine doped coatings increases to a maximum value then slowly decreases [17]. There is a small level of nitroxide present just after cure of these coatings. The concentration of nitroxide in the coating is a complex balance of formation from amine, consumption by reaction with radicals, and regeneration from aminoethers. It does not appear to be feasible to develop a model to predict the detailed time dependence of the nitroxide concentration. Several parameters can be determined from curves such as in Figure 7, including the initial net nitroxide formation rate, the nitroxide maximum and time to maximum, and the net rate of loss of nitroxide at long times. Although these parameters vary depending on the copolymer, none of the parameters appear to correlate directly either with oxidation rate or with stabilizer effectiveness. It is possible to use these parameters in conjunction with other data to derive values of the different stabilization rate constants.

The initial net nitroxide formation rates are given for the melamine and urethane coatings used in the hindered amine consumption study in Table 1 [17–19]. Also given

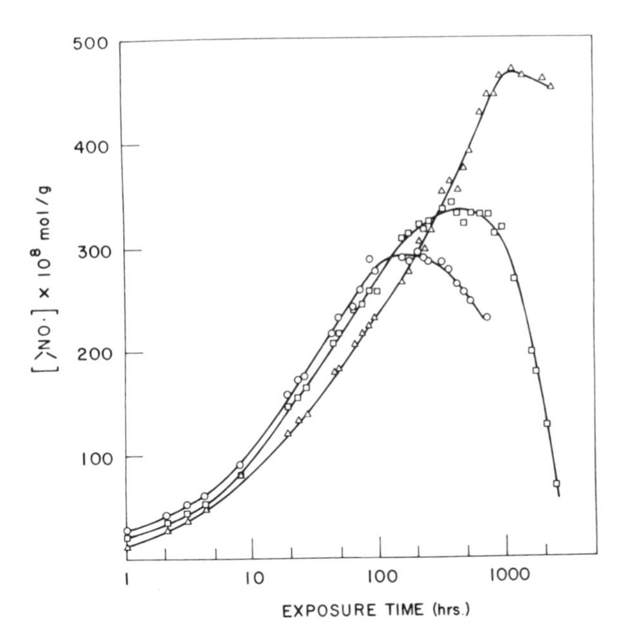

Figure 7. Nitroxide concentration versus exposure time and light intensity for an acrylic/melamine coating. Figure reprinted from [17].

are the photoinitiation rates and the initial NH consumption rates. The following points can be made: The initial net nitroxide formation rate is much smaller than either the photoinitiation rate or the initial hindered amine consumption rate; the initial net nitroxide formation rate is a strong function of coating composition and (in the case of the melamine coating) exposure humidity but is only a weak function of light intensity; higher amine consumption rates do not necessarily mean higher nitroxide formation rates. Initially, the net nitroxide formation rate (d[NO·]/dt) can be written in terms of Reactions (7), (8), (10), and (11). Nitroxide regeneration, Reaction (9),

*Table 1. Comparison on photoinitiation rates, nitroxide formation rates, and hindered amine consumption rates.**

Dew Point	PIR**	d[NO·]/dt Expt'l.	d[NO·]/dt Pred.	$-d$[NH]/dt Expt'l.	$-d$[NH]/dt Pred.
Acrylic/Urethane:					
−40°C	12	0.012	0.014	0.78	0.71
25°C	12	0.016	0.014	0.78	0.71
Acrylic/Melamine:					
−40°C	12	0.095	0.06	1.35	1.39
−40°C	7	0.092	0.12	0.95	1.04
−40°C	1.4	0.068	0.105	0.55	0.42
25°C	12	0.24	0.25	0.5	2.42†
25°C	7	0.18	0.26		
25°C	1.4	0.14	0.16		

* All data from [18].
** Photoinitiation rate of free radicals. All rates × 10^8 mole/g·min.
† Ignores hindered amine recycling reaction. See text.

can be ignored since the concentration of aminoether is small. The rates of nitroxide consumption resulting from nitroxide scavenging [Reaction (8)] and nitroxide excited state chemistry [Reactions (10) and (11)] can be calculated as discussed above and in more detail in [19]. Adding these rates to the measured net formation rates in Table 1 yield values for the rate of amine consumption, Reaction (7), and thus values for k_{NH}. The amine consumption rates calculated from the nitroxide data can be compared to the values measured in the amine consumption experiment. The dependence of nitroxide formation on light intensity can also be calculated. The calculated values are also given in Table 1. In general the agreement is quite good considering the approximations used in the analysis. The predicted value for NH consumption for the melamine coating under humid conditions is high since the recycling reaction [k_{RE}-Reaction (19)] is ignored. As expected from the arguments presented in the previous section, the rate constant k_{NH} is larger for the melamine coatings than for the urethane coating and increases with increasing humidity for the melamine coating [19].

The dependence on light intensity deserves further comment. In general, in oxidation reactions, the radical concentration depends on the square root of the light intensity. Thus it might be expected that the net nitroxide formation rate would increase with the square root of the photoinitiation rate similar to the amine consumption rate. Instead, there is very little dependence on light intensity. This result is due to the contribution of Reactions (10) and (11) to the equation for nitroxide consumption. The amount of nitroxide consumed due to nitroxide excited state chemistry is directly proportional to light intensity. Thus, the net formation rate becomes a complicated function of light intensity. The model predicts the observed behavior reasonably well.

The nitroxide concentrations in these coatings rise steadily to a maximum value and then slowly decrease. The nitroxide concentration reaches a maximum roughly when the hindered amine concentration is depleted (or has reached its plateau value). Maximum nitroxide concentrations in coatings doped with 2% by weight HALS I range from 150–400 × 10^{-8} mole/g for melamine coatings [11,17]. The values are similar though generally somewhat lower for urethane coatings [30]. In both cases, the maximum nitroxide concentration increases with decreasing light intensity. The difference observed in effectiveness between melamine and urethane coatings can not be due simply to differences in nitroxide concentration. Using Reactions (1)–(11), the following expression can be written for the maximum nitroxide concentration [19]:

$$[NO\cdot]_{max} = \frac{(k_{NH}[NH] + k_{NOY}[NOY])[YOO\cdot]}{k_{NO}[Y\cdot] + C}$$

where C is a measure of the rate of hydrogen atom abstraction by excited state nitroxide [Reactions (10) and (11)]. The values of C, k_{NO}, k_{NH}, [Y·], and [YOO·] can be determined by the methods described above. Comparing the time to reach the maximum nitroxide concentration with the hindered amine consumption experiments yields a value for [NH]. The concentration of aminoether can be estimated by mass balance (i.e., $[NOY](t) = [NH](0) - [NH](t) - [NO\cdot](t)$). Assuming that none of the rate constants change with time, it is possible to derive values for k_{NOY} from the measured values of $[NO\cdot]_{max}$ [17,30]. For the urethane coating it is found that k_{NOY} is roughly 100 times larger than k_{NH}. The predicted reduction in the oxidation chain length is large and is consistent with previously measured values [16,30]. Thus, Reactions (8) and (9) are primarily responsible for HALS stabilization in urethane coatings. The results for melamine coatings are more complicated. Holding the other rate constants fixed yields a value of k_{NOY} similar to that in the urethane coating. This results in a reduction in the oxidation chain length that is much larger than is observed experimentally [11,13]. To account for the observed HALS I effectiveness and the nitroxide concentration maximum in the melamine coating it is necessary that the value of k_{NOY} be roughly 10 times larger than k_{NH} and that the value of the ratio of k_{NO} to k_{OX} drop by about a factor of 4–5 from its initial value. The differences in the behavior of the stabilization rate constants in the melamine and urethane coatings is summarized in Table 2. Although the reaction scheme does account for the ob-

Table 2. Effect of composition and humidity on stabilization rate constants.

k_{NH}	Urethane (Dry) = Urethane (Humid) < Melamine (Dry) < Melamine (Humid). Approximate relative rates: 1:1:2:3.
k_{NO} (initial)	Urethane = Melamine, independent of humidity.
$k_{NO}(t)/k_{OX}(t)$	This ratio is independent of time in Urethanes. In melamines the ratio decreases by factor of 4–5.
k_{NOY}	Urethane > Melamine (wet) > Melamine (dry). Approximate relative rates: 5:2:1.
k_{NOY}/k_{NH}	Urethane ~100; Melamine ~10.
$\frac{k_{NH}[NH]}{k_{YH}[YH]}$	~0.02–0.04 for both urethane and melamine coatings studied.
$\frac{k_{NOY}[NOY]}{k_{YH}[YH]}$	Urethane ~1.5; Melamine ~0.4

served nitroxide concentration maximum and the observed reduction in oxidation chain length, it does not predict the observed increase in nitroxide maximum with decreasing light intensity. The reason for this behavior is unclear at present.

Explanations for the variations summarized in Table 2 are at this point speculative. The observation that k_{NOY} is much greater than k_{NH} is at odds with model compound studies for polypropylene which suggested that they were comparable [2]. The reactivity of aminoether species clearly depends both on the composition of the aminoether and attacking radical. Apparently the compositions are significantly different in coatings from the model compounds studied. The observed differences between the melamine and urethane coatings can be explained in part at least by invoking formaldehyde based chemistry. Initially, nitroxides are scavenging radicals formed by primary photochemistry. These processes are the same in melamine and urethane coatings. In the melamine coating the formation of HC·(=O) may change the balance between Reactions (2) and (8). This could account for the apparent decrease in the value of k_{NO}/k_{OX} with time. Differences in radical reactivity with nitroxide and oxygen have been reported in solution [1]. The lower value of k_{NOY} in the melamine coatings may result from a lack of an efficient way to convert NO−C(=O)H back to nitroxide. In this regard, it is interesting to note that $NOC(=O)CH_3$ is not as effective a photostabilizer in polypropylene as are species such as $NO-C_4H_9$ [6]. Reaction (19), which recycles NO−C(=O)H to NH, does not contribute directly to photostabilization since it does not consume free radicals. It produces NH groups whose stabilizing activity is not as great as NO· or other NOY groups.

So far none of the reactions mentioned irreversibly consume stabilizer. As shown in Figure 7, there is in general a slow loss of nitroxide. Surprisingly, the effectiveness of the HALS additive does not change much until the nitroxide drops to a very low value. Apparently much of the change in nitroxide concentration over this time period reflects a change in balance between the nitroxide and aminoether reactions, which may not change the overall effectiveness. When the nitroxide drops to a low level ($\sim 10 \times 10^{-8}$ mole/g), the peroxy radicals seen initially reappear, suggesting that the stabilizer is no longer effective. The exposure time required to "consume" the stabilizer depends on the coating type and on the exposure conditions. Stabilizer lifetime is shorter in coatings with high photoinitiation rates; it is longer in urethane coatings than in melamine coatings; it is roughly inversely proportional to light intensity; and it is under humid exposures longer than under dry exposures. The chemistries responsible for the depletion of stabilizer activity are unknown. Since stabilizer permanence is just as important to coating performance as stabilizer effectiveness, it is clear that coating performance should be an active area of research for years to come.

CONCLUSION

This paper summarizes studies of photodegradation in thermoset organic coatings and their stabilization by hindered amines. While much has been learned about the chemistry and kinetics of degradation and stabilization, there remain many unanswered questions. The following list is merely suggestive of the many areas of research that could be pursued:

1. Is formaldehyde based chemistry responsible for all the observed differences between melamine and urethane coatings?
2. What reactions are responsible for the slow decrease in nitroxide concentration and the ultimate loss of stabilizer effectiveness?
3. Specifically, why does humidity affect stabilizer permanence?
4. What are the structures of the aminoether species formed in thermoset coatings?
5. How does initial HALS structure influence overall stabilizer reactivity and permanence?
6. Is stabilizer migration important?
7. Are other stabilization mechanisms (e.g., quenching, specific association with or decomposition of chromophores) important in coatings?

REFERENCES

1. Brownlie, I. T. and K. U. Ingold. *Can. J. Chem.*, 45:2427 (1967).
2. Grattan, D. W., D. J. Carlsson and D. M. Wiles. *Poly. Deg. Stab.*, 1:69 (1979).
3. Felder, B., R. Schumacher and F. Sitek. *Helv. Chim. Acta*, 63:132 (1980).
4. Shilov, Yu. B. and E. T. Denisov. *Polym. Sci. USSR*, 20: 2079 (1979).
5. Carlsson, D. J., A. Garton and D. M. Wiles. In *Developments in Polymer Stabilization, Vol. I*. G. Scott, ed. London:Applied Science Pub., p. 219 (1979).
6. Kurumada, T., H. Ohsawa, T. Fujita and T. Toda. *J. Polym. Sci., Polym. Chem.*, 22:277 (1984).
7. Scott, G. *J. Photochem.*, 25:83 (1984).
8. Chan, K. H., D. J. Carlsson and D. M. Wiles. *J. Polym. Sci., Polym. Lett.*, 18:607 (1980).
9. Carlsson, D. J., K. H. Chan, J. Durmis and D. M. Wiles. *J. Polym. Sci., Polym. Chem.*, 20:575 (1982).
10. Bortolus, P., S. Dellonte, A. Faucitano and F. Gratani. *Macromolecules*, 19:2916 (1986).
11. Gerlock, J. L., D. R. Bauer and L. M. Briggs. In *Polymer Stabilization and Degradation*. P. Klemchuk, ed. ACS Symposium Series No. 280, Washington, DC, p. 119 (1985).
12. Bauer, D. R. *J. Appl. Polym. Sci.*, 27:3651 (1982).
13. Bauer, D. R. and L. M. Briggs. In *Characterization of Highly Crosslinked Polymers*. S. S. Labana and R. A. Dickie, eds. ACS Symposium Series No. 243, Washington, DC, p. 271 (1984).

14. Bauer, D. R., R. A. Dickie and J. L. Koenig. *J. Polym. Sci., Polym. Phys.*, 22:2009 (1984).
15. Bauer, D. R., R. A. Dickie and J. L. Koenig. *Ind. Eng. Chem., Prod. Res. Dev.*, 24:121 (1985).
16. Bauer, D. R., R. A. Dickie and J. L. Koenig. *Ind. Eng. Chem., Prod. Res. Dev.*, 25:289 (1986).
17. Gerlock, J. L., D. R. Bauer and L. M. Briggs. *Polym. Deg. Stab.*, 14:53 (1986).
18. Gerlock, J. L., T. Riley and D. R. Bauer. *Polym. Deg. Stab.*, 14:73 (1986).
19. Bauer, D. R. and J. L. Gerlock. *Polym. Deg. Stab.*, 14:97 (1986).
20. Gerlock, J. L., D. R. Bauer and L. M. Briggs. In *Characterization of Highly Crosslinked Polymers*. S. S. Labana and R. A. Dickie, eds. ACS Symposium Series No. 243, Washington, DC, p. 285 (1984).
21. Gerlock, J. L., H. van Oene and D. R. Bauer. *Euro. Polym. J.*, 19:11 (1983).
22. Gerlock, J. L. *J. Anal. Chem.*, 55:1520 (1983).
23. Gerlock, J. L. and D. R. Bauer. *J. Polym. Sci., Polym. Lett.*, 22:447 (1984).
24. Gerlock, J. L., D. R. Bauer, L. M. Briggs and R. A. Dickie. *J. Coat. Technol.*, 57(722):37 (1985).
25. Emanuel, N. M., E. T. Denisov and Z. K. Maizus. *Liquid Phase Oxidation of Hydrocarbons*. New York:Plenum Press (1967).
26. Ingold, K. U. *Chem. Rev.*, 61:563 (1961).
27. Bogatryeva, A. I. and A. L. Buchachenko. *Kinetics and Catalysis*, 12:1226 (1971).
28. Bauer, D. R., L. M. Briggs and J. L. Gerlock. *J. Polym. Sci., Polym. Phys.*, 24:1651 (1986).
29. Keana, J. F. W., R. Dinerstein and F. Baitis. *J. Org. Chem.*, 36:209 (1971).
30. Bauer, D. R., M. J. Dean and J. L. Gerlock. *Ind. Eng. Chem.* (submitted).
31. Berge, A., B. Kvaeven and J. Ugelstad. *Euro. Polym. J.*, 6:981 (1970).
32. Gerlock, J. L., M. J. Dean, T. J. Korniski and D. R. Bauer. *Ind. Eng. Chem., Prod. Res. Dev.*, 25:446 (1986).
33. Moore, M. L. In *Organic Reactions, Vol. 5*. R. Adams, ed. New York:John Wiley & Sons, chapter 7, p. 307 (1949).

N. C. Billingham[1]
J. W. Burdon[1]
I. W. Kaluska[2]
E. S. O'Keefe[1]
E. T. H. Then[1]

Chemiluminescence from Oxidative Degradation of Polymers

ABSTRACT

Studies of the ultra-weak chemiluminescence which accompanies polymer oxidation are shown to provide an extremely sensitive way of monitoring degradation. It is shown that luminescence can be observed not only during polymer oxidation but also when samples are heated in an inert atmosphere. Both types of luminescence have been studied for the oxidation of natural rubber and Nylon 6.6 and for the electron beam irradiation of Nylon 6.6 and of polyethylene. In unirradiated samples the luminescence can be correlated with the peroxidation of the polymer although rather different mechanisms of luminescence are involved in rubbers and in Nylon. The luminescence of irradiated polymers is shown to be more complex and probably involves more than one mechanism.

KEY WORDS

Chemiluminescence, oxyluminescence, polymer oxidation, polymer irradiation, natural rubber, Nylon 6.6, polyethylene.

INTRODUCTION

After many years of research there is general agreement about the chemistry of the reactions which lead to peroxidation, chain scission and embrittlement in polymers and many effective stabilizers have been developed. There remains, however, a number of problems. In particular, the prediction of polymer lifetimes is still a challenge, based mostly on extrapolations of accelerated aging tests. The problem is that although we know what reactions are involved in the degradation sequence, almost all of our knowledge of their kinetics is derived from studies of reactions in the liquid phase and we know rather little of how the polymer matrix affects the reactions. The difficulty is that the reaction rates are very low, so that the concentrations of the intermediate free-radicals are too small for measurement by any of the conventional methods, as are the concentrations of stable intermediates. Although progress has been made, it has mostly involved studies of γ-initiated oxidation where the radical concentrations are much higher than normal. There is a clear and established need for more sensitive techniques for probing oxidation. One possibility is the measurement of the associated light emission.

The emission of very low levels of light ("oxyluminescence"), which accompanies the oxidation of all organic materials, has been known and well documented for many years [1,2]. The first application of oxyluminescence measurements to polymers was reported in 1961 by Ashby [3], who showed that many polymers emit light very weakly when they are heated to around 180°C in air and that the emission is associated with oxidation. He showed that luminescence is most intense from polymers containing amide groups (Nylon and polyurethanes) and least intense from hydrocarbon polymers. The total amount of light emitted was found to be proportional to the concentration of carbonyl groups developed in the polymer, although at least 10^4 carbonyl groups were estimated to be formed per photon emitted. Ashby also showed that the luminescence is suppressed or reduced by antioxidants and proposed its measurement as a technique for evaluation of antioxidants. Similar studies of polymers were reported by Schard and Russell [4,5], who also demonstrated the connection between luminescence and oxidation. These early studies were limited by the low signal to noise ratio of analogue electrometers so that oxidation could be observed only at high temperatures, where other techniques are easier to use and provide more direct results. In recent years, the development of fast single-photon counters has enabled low light levels to be measured with much greater sensitivity and there has been new interest in oxyluminescence.

One problem in the interpretation of luminescence data has been that there is no general agreement on the origin of the luminescence. The requirements for a reaction to lead to luminescence are quite restrictive, in that the product must be formed in an excited state, with sufficient energy to emit a detectable photon and a finite prob-

[1]School of Chemistry and Molecular Sciences, University of Sussex, Brighton, BN1 9QJ, England.
[2]Institute for Nuclear Chemistry and Technology, Warsaw, Poland.

ability of decay to the ground state via light emission, rather than radiationless deactivation. The intensity of oxyluminescence is too low for it to be measurable when dispersed by a high-resolution monochromator, but simple experiments with filters have suggested that the emission from Nylon has a peak above 350 nm, similar to the phosphorescence of an excited carbonyl group [6]. This emission requires the luminescent reaction to have an energy yield of at least 300 kJ mol^{-1}, which implies either a highly exothermic reaction or one with a high activation energy. It is generally accepted that the emission comes from phosphorescence of an excited ketone, but opinions differ on the origin of the ketone. Many workers [2,5,6] have proposed that it is produced in the termination reaction of two secondary alkyl-peroxy radicals via the mechanism:

$$RO_2^{\cdot} + R_2CH{-}O_2^{\cdot} \longrightarrow ROH + {}^1O_2 + R_2C{=}O$$

This model seems to work well for liquid phase oxidation and has been used widely. For example, Mendenhall et al. [7] claimed that low-temperature luminescence measurements could be interpreted by kinetic models based on this mechanism and could be useful for a wide variety of substrates, including aviation fuels and food oils as well as polymers.

In contrast, other workers [8] have suggested that the luminescence is produced by the decomposition of the polymer hydroperoxide. Simple unimolecular decomposition is not sufficiently exoenergetic to allow luminescence and it must arise either from a molecular rearrangement reaction [9] or from a two stage process in which primary radicals, produced by decomposition of peroxides, disproportionate within the cage in which they are produced [10]:

$$\text{H-C-OOH} \longrightarrow [\text{H-C-O}^{\cdot} + {}^{\cdot}\text{OH}] \longrightarrow \text{C=O} + H_2O$$

The distinction between these two mechanisms is essentially semantic but it should be noted that neither pathway requires oxygen to be present.

Zlatkevich [11] has used this model to interpret measurements of oxyluminescence and claims to be able to make lifetime predictions. Recently, Mendenhall and Quinga [13] pointed out that the disproportionation of alkoxy radicals can give luminescence with high quantum efficiency and suggested that this reaction might play a part in the oxyluminescence from polymers. This mechanism is very similar to that proposed for peroxide decomposition, with the exception that the radicals involved are not primary radicals. It is important to recognise that the rates of initiation and termination are expected to be equal in the steady-state conditions used in most studies so that mechanisms involving peroxide decomposition (initiation) may not be kinetically distinguishable from those involving radical recombination (termination).

Some elegant work has been reported by George, who has reviewed his contribution [6]. He assumes that luminescence arises from chain termination and uses this model to interpret the results of experiments (pioneered by Shlyapintokh et al. [1] for liquids), in which the sample is perturbed by a sudden change in either temperature or atmosphere and its return to steady-state conditions is monitored. He interpreted results for Nylon 6.6 in terms of the relative rate constants for mutual termination of alkyl (in the absence of oxygen) and alkyl-peroxy (upon oxygen admission) reactions, assuming a constant rate of initiation from the decomposition of adventitious peroxides. The work was all carried out on commercial Nylon fibres, which had been extracted to remove stabilizers. Complete removal of the stabilizer from a highly drawn fibre is almost impossible and radical mobilities in highly drawn fibres may be very much more restricted than in unstressed materials so that there is some query as to how far these conclusions can be generalised to other polymers or even to unoriented Nylon.

Clearly if chemiluminescence measurements on polymers are to be placed on a quantitative basis, it is desirable to have a better understanding of the mechanisms of luminescence and how they relate to the overall oxidation sequence. With this desire in mind we have been studying luminescence from a variety of polymers. We here report some of the phenomena which can be observed in three different systems of increasing complexity—the autoxidation of natural rubber, the autoxidation of Nylon 6.6 and the electron beam irradiation of Nylon and polyethylene.

EXPERIMENTAL

Photon Counting

Oxyluminescence emission from polymers under conditions approximating end use is extremely weak, often less than 100 photons per second reaching the detector in a typical experimental geometry. This observation means that very sensitive detection is needed. We infer that the use of a cooled photomultiplier as detector and an efficient pulse analyser is needed to carry out the photon counting. The instrument which we use was developed in our own laboratory using a micro-computer to carry out the photon counting and also to control the temperature of the sample and the gas flow over it. Full details of the instrument have been given elsewhere [14]. The sample is usually in the form of a thin film with dimensions of around 2.5 cm square. Whilst there is no objection to thick samples, thin ones give better temperature control and reduce the risk of diffusion control effects. We have recently modified the equipment to allow monitoring of luminescence response following irradiation of the sample with UV light and to allow measurements of the emission spectra of luminescence using a computer controlled linear interference filter.

Polymer Samples

Natural rubber was a sample of Sri-Lankan pale crepe rubber, purified by repeated reprecipitations to remove proteins and natural antioxidants. Samples were prepared for luminescence measurements by casting from toluene solution onto clean aluminum pans. Nylon 6.6 was additive-free, melt extruded, unoriented film. Polyethylene was a low density material extruded as 100 μm film without stabilizers. All films were washed in alcohol-free chloroform to remove any surface contamination before being used.

Other Techniques

Peroxide contents in natural rubber were determined by iodometric analysis using the method of Carlsson and Wiles [15]. In Nylon the measurement was made by using the oxidation of Fe(II) in hexafluoroisopropanol solution [16]. Electron beam irradiations were carried out at room temperature in a LINAC accelerator at a dose rate of 2.68 kGy min^{-1}.

RESULTS AND DICUSSION

Natural Rubber

Natural rubber is one of the simplest systems for oxyluminescence study. It combines a reasonably high rate of oxidation with detectable chemiluminescence. The polymer is well above its T_g at the temperatures at which it oxidises so that the oxidation can reasonably be modelled as a liquid-phase reaction. Studies of rubbers have been reported by Mendenhall et al. [17] and by Mayo and Davenport [18]. Figure 1 shows the luminescence curve for the steady-state oxidation of a sample of natural rubber. The close parallel with the expected oxidation curve is obvious, suggesting that the intensity of luminescence is proportional to the rate of the oxidation reaction as required by both luminescence mechanisms.

If the sample is heated in a series of stages, it is possible to obtain an Arrhenius plot of the luminescence intensity and this gives an activation energy of the order of 100 kJ mol^{-1}, comparable with values obtained from oxygen absorption and peroxide formation measurements.

A rather more revealing observation is the heating of a rubber sample in nitrogen, after thorough flushing to remove dissolved oxygen. A luminescence peak is produced which can easily be measured at high sensitivity and which decays away to a very low level over a period of minutes to hours, depending on the temperature. Figure 2 shows such a peak on an expanded scale.

The area of this peak represents the total number of photons that are emitted by the sample on heating. For convenience we call this the total luminous intensity (TLI) of the sample. We find that the TLI is directly proportional to the weight of the sample, from which we infer that luminescence is from the whole of the sample rather than a surface layer. The peak has many characteristics suggestive of an association with the peroxide content of the sample. Thus it appears only on the first heating of the sample; if the sample is cooled in nitrogen then reheated there is no further luminescence. Brief exposure to air does not regenerate the peak on reheating but if the sample is allowed to age in air after the experiment then

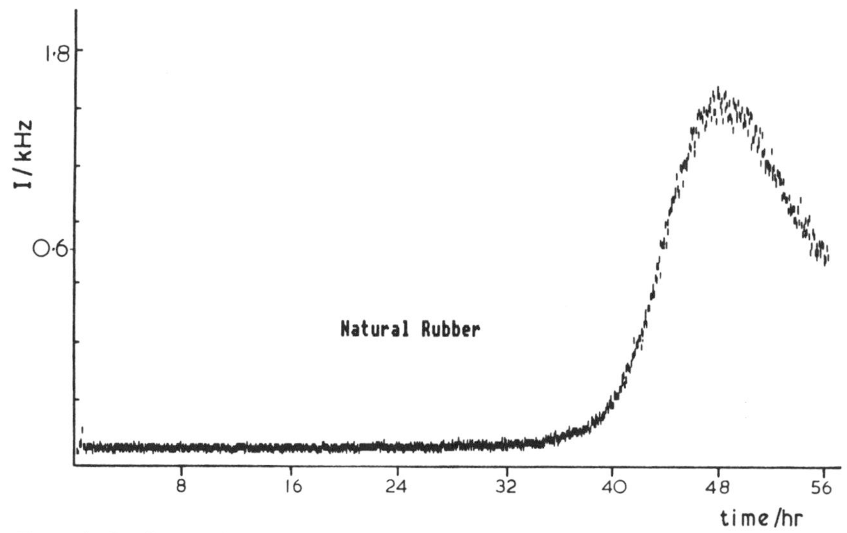

Figure 1. Continuous chemiluminescence from a thin film of natural rubber in oxygen at 80°C.

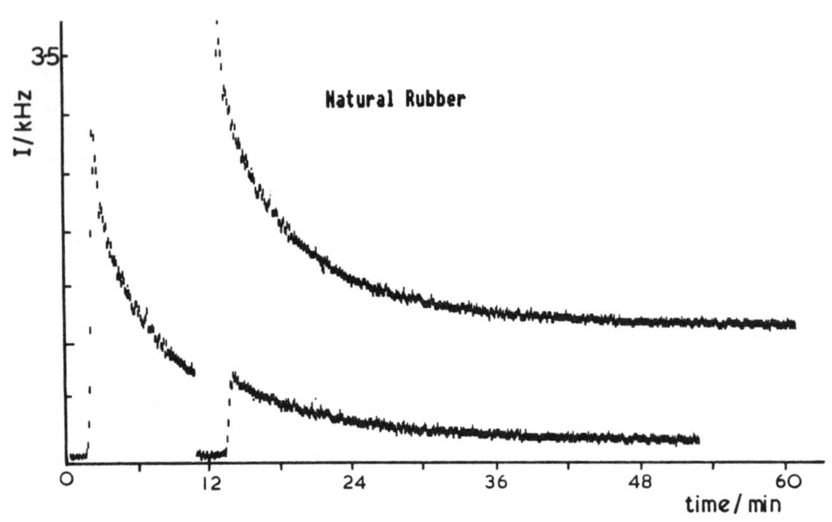

Figure 2. Chemiluminescence from a sample of natural rubber heated at 100°C in nitrogen. Lower curve shows effect of cooling sample to room temperature and reheating. Upper curve is duplicate experiment without cooling but on a shifted scale.

the peak does reappear on heating. If the temperature is lowered during the observation of the peak then the intensity of luminescence decreases to become immeasurably small. On restoring the temperature we return to the original peak as is shown in Figure 2. This observation shows that the reaction producing the luminescence is continuous and not the decay of a transient population of reactive species. Finally, treatment of the rubber with peroxide-destroying agents, especially SO_2, causes the initial peak to be reduced markedly in intensity. All of these results are consistent with the proposal that the initial peak can be directly linked to the peroxide content of the rubber.

To explore this relation further we carried out a series of experiments in which samples of the rubber were oxidised for varying periods in air then cooled, flushed with nitrogen, and reheated in nitrogen. Figure 3 shows that there is a linear relationship of the TLI of a polymer sample to the steady luminescence from the same sample at the point at which the oxidation was stopped.

In another series of experiments we oxidised a series of rubber samples in air for varying periods and measured their peroxide content by iodometry and their TLI. Figure 4 shows the evolution of both peroxide content and TLI at 120°C and Figure 5 shows the relationship of the peroxide content and the luminescence for experiments

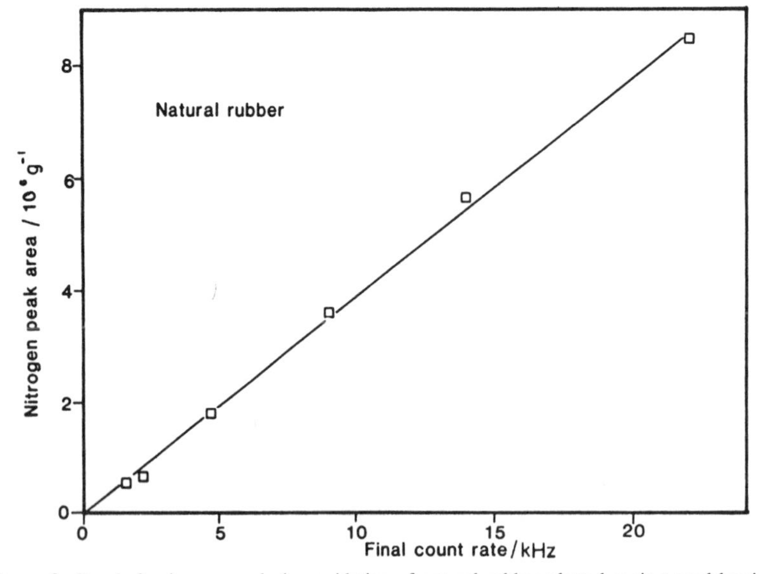

Figure 3. Steady luminescence during oxidation of natural rubber plotted against total luminescence on heating under nitrogen.

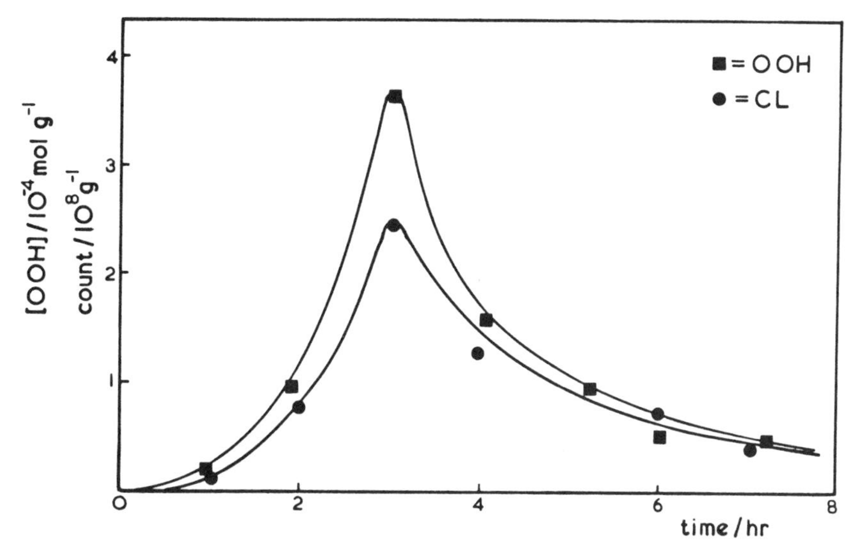

Figure 4. Formation of peroxides (■) and evolution of total chemiluminescence (●) during air oxidation of natural rubber at 120°C.

at 120 and 140°C. The slope of these graphs corresponds to the emission of about 6 × 10^{11} photons per mole of OOH groups, or one photon emission for every 10^{12} OOH groups decomposed. Even allowing for the fact that the quantum efficiency for the emission is likely to be of the order of 10^{-8}, one still infers that luminescence is a rare phenomenon.

We conclude that the luminescence in rubber samples is associated with the decomposition of a very small fraction of the hydroperoxides in the sample, so that luminescence in air monitors the rate of initiation of the degradation whilst luminescence in nitrogen measures the peroxide content. Whether the luminescence arises directly from the reactions of the peroxides or via recombination of primary radicals remains uncertain. The important feature is the fact that the iodometric and oxygen absorption methods are operating close to their detection limits under the conditions of our study, whereas the luminescence method has reserves of sensitivity. Thus we have been able to detect luminescence from stabilized, carbon-black filled rubbers and further studies of stabilized samples are in progress.

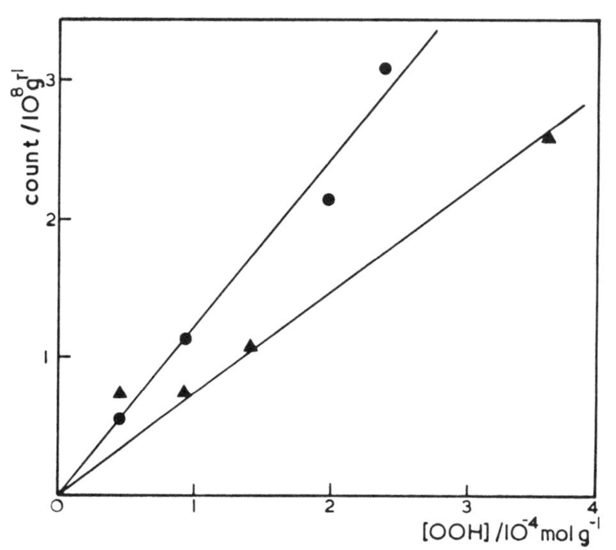

Figure 5. Dependence of total luminescence under nitrogen on peroxide content for oxidation of natural rubber at 120°C (●) and 140°C (▲).

Nylon 6.6

Polyamides are very much more stable to oxidative degradation than are rubbers. In addition they are typically semi-crystalline, with glass transition temperatures above room temperature. We may therefore expect that radicals produced in a Nylon sample will have much more restricted mobilities than in rubbers, which might well affect the chemiluminescence behaviour. For our studies we have used unstabilized, unoriented films of Nylon 6.6. We have previously reported details of some of these studies [19].

The luminescence curve obtained from samples of Nylon film heated continuously in air shows a slowly autoaccelerating oxidation curve with essentially no induction period and an activation energy of around 80 kJ mol^{-1}. This energy is lower than would be expected for the unimolecular decomposition of an isolated hydroperoxide group (around 140 kJ mol^{-1}), suggesting that initiation is either by unimolecular decomposition catalysed by impurities, or by bimolecular decomposition of associated peroxides.

Because the degradation is slow, continuous steady-state measurements are less informative for Nylon than for rubber—although the pattern is similar. However,

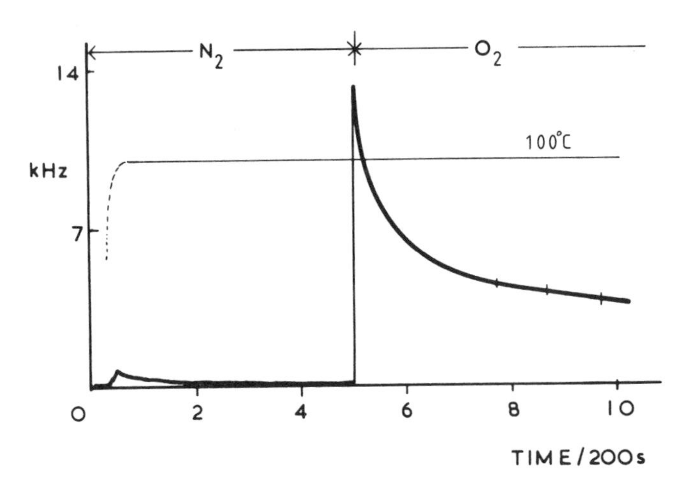

Figure 6. Chemiluminescence response of a sample of Nylon 6.6, heated under nitrogen to 100°C, followed by admission of oxygen after 500 s.

like rubber, Nylon shows a very weak luminescence emission peak, typically a few hundred Hz, on heating in nitrogen. When this peak is over and oxygen is admitted to the cell, the response is very different from that of rubber, in that there is a near instantaneous increase in intensity to a few kHz followed by a decay back to a lower intensity as shown in Figure 6. The peak and decay in intensity on oxygen admission have been reported before [6] but the peak in nitrogen had not been reported before our work.

The initial, weak peak observed on heating can be expanded and studied. We find that it has many of the same features as the peak seen in rubbers although the decay is very much more rapid so that it is harder to study. It occurs only on first heating and its shape and intensity are apparently not very sensitive to traces of oxygen; identical results are obtained if the cell is flushed with nitrogen for 30 min or evacuated to 10^{-3} torr for 24 hr prior to heating. However, it reappears if the sample is briefly exposed to oxygen at room temperature then reheated in nitrogen. Its area is reduced by 75% if the sample is exposed to SO_2 for 12 hr before heating. These results strongly suggest that the luminescence is associated with decomposition of peroxides in the sample.

If the nitrogen peak comes from a simple unimolecular decomposition process, we would expect to find that the part of the peak observed at constant temperature would follow first-order kinetics and this is not so. In contrast,

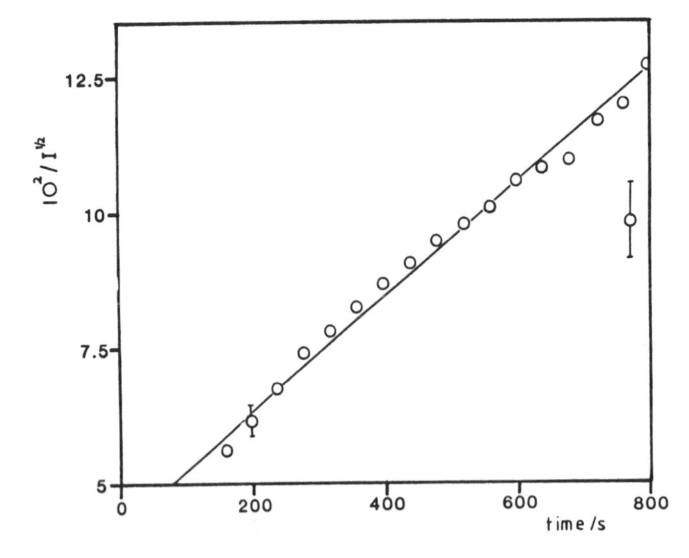

Figure 7. Test of second-order kinetic model for the decay of luminescence under nitrogen in Nylon 6.6 at 100°C.

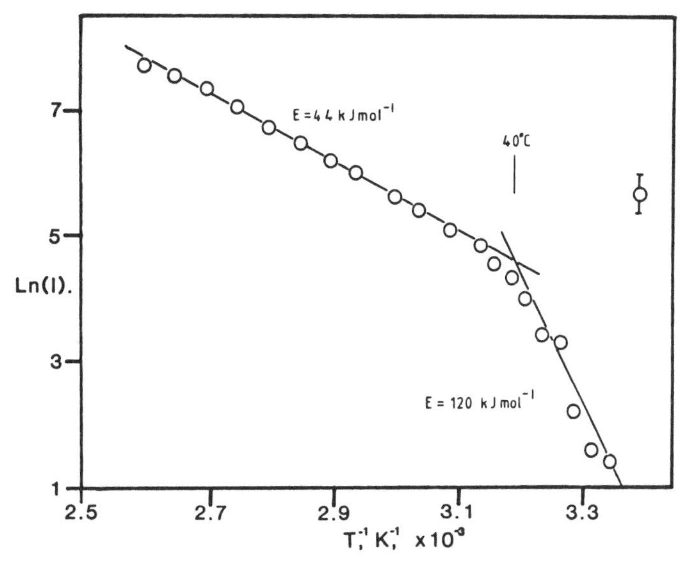

Figure 8. Arrhenius plot for the luminescence of Nylon 6.6 during heating in nitrogen.

if the observed decay in intensity resulted from a relatively slow recombination of radicals then a plot of $1/I^{0.5}$ against time should be linear; the ratio of its intercept and slope give the half-life of the radicals. Figure 7 shows that this relation fits the data well for at least 80% of the decay, the slope of the line corresponding to a half-life of 6.5 min at 100°C, too short for a peroxide decomposition reaction (note that because this is a second-order reaction the half-life depends on the initial concentration as well as the rate constant).

It is possible to slow down the rate of heating of the sample and explore the temperature dependence of the luminescence and Figure 8 shows the Arrhenius plot thus obtained. Up to 40°C, the activation energy is around 120 kJ mol^{-1} but it quickly decreases to a much lower value of 44 kJ mol^{-1} above this temperature. The lower value seen over most of the temperature range is much too low for a unimolecular decomposition of a peroxide, but is more consistent with recombination of mobile species. It is very tempting to speculate that the change in activation energy is associated with increasing mobility as the glass transition temperature of the polymer is approached.

We believe that the luminescence of Nylon in nitrogen is caused by the decomposition of a small fraction of the total hydroperoxides in the sample but results from recombination of relatively long-lived oxygenated product radicals. This peak does not appear to be due to traces of dissolved oxygen. It is a sensitive monitor of the peroxide content of the polymer and of its ageing history, and is observable and measurable well before any OOH groups can be detected in the film by IR or chemical methods.

When oxygen is admitted to the sample cell after a period of heating in nitrogen, there is a very sharp step increase in the luminescence. The height of this step initially increases with the time of heating under nitrogen but becomes constant after long heating times, as shown in Figure 9. Thus, despite the evidence of near complete decomposition of the hydroperoxides in the sample, this height is the same if the polymer is maintained for up to seven days in nitrogen after completion of the first peak and before admission of oxygen.

These results imply that the species responsible for luminescence are produced in the first stage heating and can be "stored" in the polymer for long periods. In particular a stationary concentration of radicals, as proposed by George [6], would imply, that an initially very low concentration of peroxides in the sample decomposes at a rate high enough to maintain a steady concentration of radicals, without any change in the peroxide concentration over a period of days at 100°C. This is most improbable and we prefer to propose that all of the hydroperoxides present initially are decomposed by heating and that, in the absence of oxygen, hydrogen abstraction or chain scission converts all of those primary radicals which do not terminate into alkyl radicals. At least some fraction of these radicals remain stable in the polymer and do not recombine at a significant rate. (It is not clear whether radicals are in any real sense "trapped" in the polymer or simply recombine very slowly because of their low concentration.) When oxygen is admitted, all of these alkyl radicals are converted rapidly into alkylperoxy radicals. These radicals terminate, with luminescence, both because of their inherently higher mobility and because they move through the polymer by propagating the oxidation chain. The very large difference between the lifetime of the alkyl radicals, which appears to be indefinite, and that of alkyl-peroxy radicals, a few

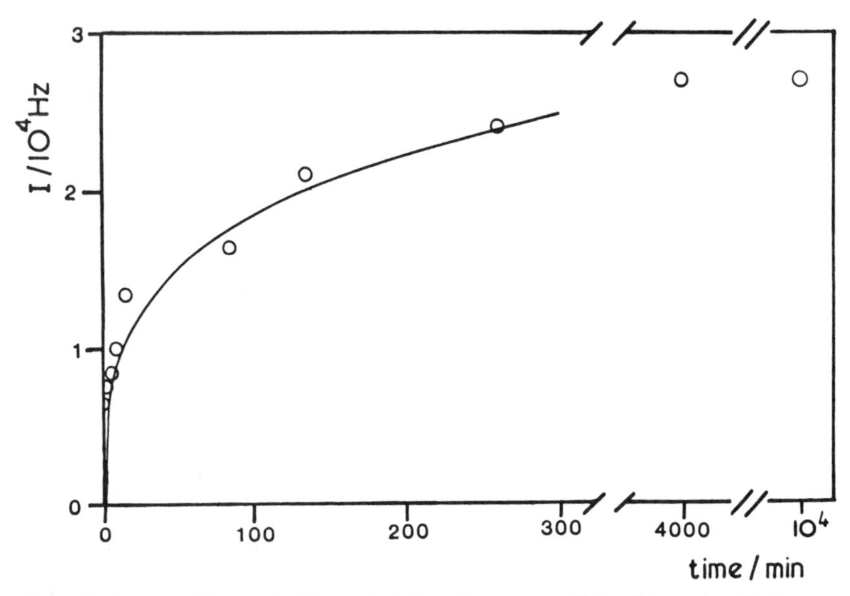

Figure 9. Increase of the step height on admission of oxygen to a Nylon 6.6 sample with the period of initial heating under nitrogen at 100°C.

minutes, suggests that the rate of termination of the alkylperoxy radicals is mainly determined by their rate of movement through the polymer matrix by chain propagation.

Irradiated Polymers

Irradiation of polymers is technologically important, both as a means of modifying properties, typically by cross-linking, or as a means of sterilization, as in medical applications. It is well established that both electron-beam and γ-radiation will lead to the formation of hydroperoxides and free-radicals in polymers and that both may lead to serious problems of post-irradiation oxidation. There are thus two reasons for interest in the chemiluminescence processes accompanying irradiation: the hope of gaining new insight into the luminescence processes and the hope of developing new methods for evaluation of the damage done to the polymer by irradiation. With these ideas in mind we have looked at the effects of electron-beam irradiation on Nylon and low-density polyethylene.

Electron beam irradiation of polyethylene is widely used to induce cross-linking. It is well established [20–22] that a polymer irradiated at 77 K and then heated to room temperature emits light in a series of peaks at temperatures which can be identified with transitions in the polymer. This thermoluminescence has been attributed to the relaxation of electrons trapped in the polymer matrix but is essentially complete before the polymer reaches room temperature. Mihalcea et al. [23] suggested that the total light emitted during heating of a sample of polyethylene irradiated at low temperatures decreases with dose in the range 0.1–10 kGy. They suggested that this method could be used to determine degree of cross-linking but did not comment on the mechanism of the luminescence. Fisher [24] reported that polyethylene irradiated at room temperature shows a peak in luminescence when heated. The TLI increased with dose up to 20 kGy and decreased thereafter. In the region above 20 kGy the decreasing light intensity was paralleled by a decrease in the ability of the polymer to

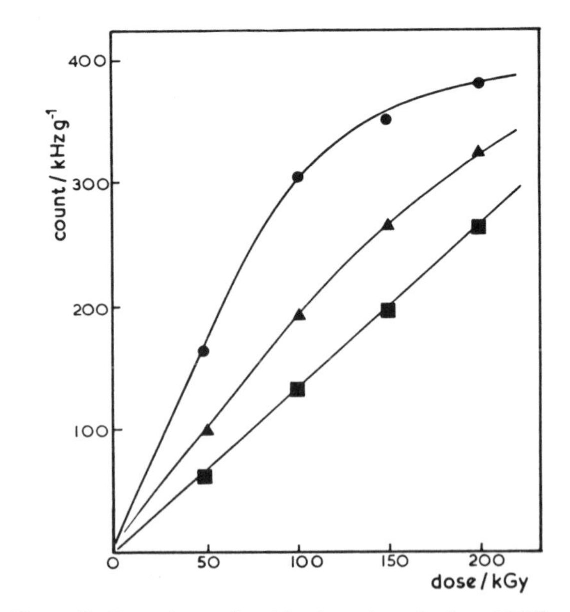

Figure 10. Dependence of total luminous intensity from LDPE on the electron beam dose. ● = immediately after irradiation; ▲ = after 24 hr in air at 25°C; ■ = after 28 days in air at 25°C.

destroy the stable free-radical DPPH and it was suggested that the luminescence is associated with peroxy radicals.

We also have found that irradiated polyethylene shows a chemiluminescence peak when it is heated from room temperature to 100°C. This peak is weaker than a typical thermoluminescence peak but more intense than that found for unirradiated polymer. In contrast with other reports we find that the TLI increases with dose, as shown in Figure 10. The peak cannot be used to estimate cross-linking since it is not storage-stable, decreasing by around 50% if the samples are stored at room temperature for 4 weeks after irradiation. Heating to 80°C causes 50% loss in area in about 3 hr. The initial relationship of TLI to dose is curved and becomes more linear as the sample ages. ESR studies show that there is a measurable population of both alkyl and alkyl-peroxy radicals in the polymer immediately after irradiation but that they decay to an immeasurably low level in a period of hours at room temperature, much faster than the decline in the luminescence peak. Furthermore, the luminescence decreases only by 13% if the polymer is allowed to swell in hexane for 12 hr then dried. Both of these results suggest that luminescence is not simply due to recombination of the radicals produced by irradiation. Although this would suggest that the peroxides formed on irradiation might be responsible for the light emission, peroxide analysis does not support this suggestion; samples irradiated to 100 kGy showed the same peroxide concentration, as measured by iodometry, both immediately after irradiation and after a further 4 months at room temperature in air.

Experiments with irradiated Nylon 6.6 have revealed an even more complex picture. As with polyethylene, irradiation leads to samples which luminesce strongly on heating in nitrogen. However, as is shown in Figure 11, the TLI passes through a maximum at a low dose then falls, so that a sample irradiated to 200 kGy is not much more luminescent than an unirradiated sample.

ESR measurements again show that the concentration of alkyl and alkyl-peroxy radicals in the sample falls rapidly in air at room temperature. As with polyethylene, the luminescence process is dependent on the treatment of the sample; in particular, samples irradiated to low doses lose most of their luminescence on standing in air or on swelling, whereas high dose samples are virtually unaffected by these treatments. Virtually all of the TLI is suppressed by exposure to SO_2. At the high dose rates involved in these experiments the oxidation will be extremely diffusion controlled and we expect all of the dissolved oxygen in the film to be consumed within seconds. We might then assume that further irradiation would lead to destruction of peroxides which could explain the decay in luminescence. Unfortunately, peroxide analysis shows a continuous increase in peroxide concentration with irradiation time. At this stage we believe that the luminescence in these irradiated polymers must arise from more than one mechanism and further experiments are in progress to try to get a more detailed picture.

CONCLUSIONS

Oxyluminescence measurements are an exceedingly sensitive way of studying the degradation processes in polymers. In a simple polymer system, such as natural rubber, the emission correlates very well with other measures of oxidation and there seems to be a very real pros-

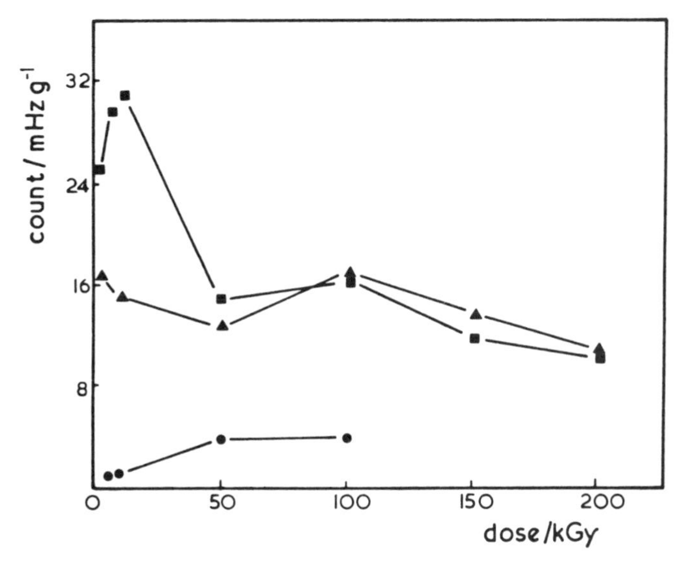

Figure 11. TLI from irradiated Nylon 6.6 as a function of electron beam dose. ■ = immediately after irradiation; ▲ = after 48 hr at 25°C; ● = after 12 hr exposure to sulphur dioxide.

pect of taking advantage of the high sensitivity of the method to develop a useful aging test for low-temperature application. In a polymer like Nylon, with its more rigid and semi-crystalline matrix the luminescence is more complex. We believe that it can be understood at least qualitatively in terms of the decomposition of polymer hydroperoxides and the constraints imposed by the polymer on radical mobility. Correlations with other ageing tests are more difficult because luminescence measurements show measurable effects well before any degradation is apparent by other methods. However there does seem to be a real prospect of developing a useful correlation. Experiments on irradiated polymers have shown greater complexity than has previously been implied. Further work is needed to clarify the processes involved.

ACKNOWLEDGEMENTS

The support of this work by grants from the Royal Aircraft Establishment and the Admiralty Research Establishment is gratefully acknowledged. We are also grateful to Dr. D. H. MacKerron, formerly of DuPont Canada Inc., for the Nylon film samples and to Raychem Ltd. for electron-beam irradiations.

REFERENCES

1. Shlyapintokh, V. Ya. et al. *Chemiluminescence Techniques in Chemical Reactions*. New York:Consultants Bureau (1968).
2. Vasil'ev, R. F. *Russian Chem. Revs.*, 39:529 (1970).
3. Ashby, G. E. *J. Polymer Sci.*, 50:99 (1961).
4. Schard, M. P. and C. A. Russell. *J. Appl. Polymer Sci.*, 8: 985 (1964).
5. Schard, M. P. and C. A. Russell. *J. Appl. Polymer Sci.*, 8: 997 (1964).
6. George, G. A. *Dev. Polymer Deg.*, 3:179 (1981).
7. Mendenhall, G. D. *Angew. Chem. Int. Ed.*, 16:225 (1977).
8. Reich, L. and S. S. Stivala. *Makromol. Chem.*, 103:74 (1967).
9. Lloyd, R. A. *Trans. Far. Soc.*, 61:2182 (1965).
10. Vasil'ev, R. F. *Makromol. Chem.*, 126:231 (1969).
11. Zlatkevich, L. *J. Polymer Sci.*, 21:571 (1983).
12. Zlatkevich, L. *Poly. Prepr.*, 25:81 (1984).
13. Quinga, E. M. Y. and G. D. Mendenhall. *J. Amer. Chem. Soc.*, 105:6520 (1983).
14. O'Keefe, E. S. and N. C. Billingham. *Polymer Deg. Stab.*, 10:1347 (1985).
15. Carlsson, D. J. and D. W. Wiles. *Macromolecules*, 2:598 (1969).
16. Marchal, J. *Radiat. Phys. Chem.*, 16:27 (1980).
17. Mendenhall, G. D., R. A. Nathan and M. A. Golub. In *Applications of Polymer Spectroscopy*. E. G. Brame, Jr., ed. NY:Acad. Press, p. 101 (1978).
18. Davenport, J. E. and F. R. Mayo. *Polym. Prepr.*, 25:79 (1984).
19. Billingham, N. C. and E. S. O'Keefe. 6th International Conference, Lucerne (1985).
20. Charlesby, A. and R. H. Partridge. *Proc. Roy. Soc., Lond.*, A271:170, 188 (1963); A283:312 (1965).
21. Markiewicz, A. and R. J. Fleming. *J. Polymer Sci., Phys.*, 24:1713 (1986).
22. Mozisek, M. *Int. J. Appl. Radiation*, 21:11 (1970).
23. Mihalcea, I., S. Jipa and T. Vladimirescu. *Radiochem. Radioanl. Lett.*, 43:19 (1980).
24. Fisher, W. K. *J. Indust. Irrad. Technol.*, 3:167 (1985).

D. BRAUN[1]
B. BÖHRINGER[1]
W. KNOLL[1]
W. MAO[1]

Thermal Degradation of Poly(Vinyl Chloride) Blends

ABSTRACT

Poly(vinyl chloride) was mixed with various poly(methacrylate)s by combined precipitation from common solutions. The thermal stability of the samples was measured at 180°C under nitrogen, and the HCl evolved was detected by conductometry.

UV-Vis-spectra of degraded samples were measured to investigate the influence of the poly(methacrylate)s onto the lengths of polyenes formed during the degradation of poly(vinyl chloride).

The experiments show that the nature of the ester group is the dominating factor for the thermal stability of poly(vinyl chloride) in these blends. Poly(n-butylmethacrylate) exhibits the best stabilisation for poly(vinyl chloride) in this series.

Stabilisation experiments with dibutyltin-bis-iso-octyl-thioglycollate show a costabilising effect of the poly(methacrylate)s.

KEY WORDS

PVC, PMMA, Poly(methacrylate)s, polymer blends, thermal stability, dehydrochlorination, polyenes, stabilisation.

INTRODUCTION

Poly(vinyl chloride) (PVC) is one of the most important thermoplastics. A limitation for its application is its rather poor thermal stability. In practise, this problem is overcome by the use of stabilisers. The thermal stability of PVC has been thoroughly investigated (see for example [1–3] and references cited therein).

In recent years PVC has gained interest as a component in polymer blends due to its miscibility with various polymers containing ester groups [4]. Poly(methacrylate)s are one group of polymers, whose miscibility with PVC has been investigated [5–7]. These investigations show that the miscibility of poly(methacrylate)s with PVC varies with the ester side chain; i.e., the longer the side chain, the less is the miscibility. Poly(methacrylate)s, especially poly(methylmethacrylate) (PMMA), are used as processing aids for PVC [8]. Therefore, detailed knowledge about the influence of a second polymer on the thermal stability of PVC would be useful.

Very few experiments in this field have been carried out. In 1970 McNeill and Neil [9] investigated the thermal stability of PVC/PMMA mixtures. In recent investigations [10], a dependence of the thermal stability of PVC on the miscibility with polyesters as polycaprolactone and polyethylene-adipate is discussed.

In this work the length of the ester side chain in poly(methacrylate)s was varied to change the miscibility with PVC. In detail, PMMA, poly(ethylmethacrylate) (PEMA), poly(n-butylmethacrylate) (PnBMA), poly(t-butylmethacrylate) (PtBMA), poly(hexylmethacrylate) (PHMA) and poly(decylmethacrylate) (PDMA) were used.

With the PVC/PMMA and the PVC/PnBMA blends, stabilisation experiments have also been carried out. These experiments should give hints whether the effect of stabilisers for PVC may be influenced by the presence of a second polymer.

EXPERIMENTAL

The PVC used in this work was Vestolit 6554 (Hüls AG, Marl). It was precipitated with tetrahydrofurane (THF)/methanol before use and dried in vacuo.

The poly(methacrylate)s were laboratory products, kindly supplied by Röhm GmbH (Darmstadt) through the courtesy of Drs. Wunderlich and Albrecht. These polymers were precipitated from methylene chloride/methanol and dried in vacuo.

The characteristic data of the polymers are given in Table 1.

All solvents were purified in the usual manner. THF was refluxed 8 hr with calciumhydride under argon and then 8 hr with potassium also in an argon atmosphere. This procedure was necessary to remove peroxides.

[1]Deutsches Kunststoff-Institut, D-6100 Darmstadt, Federal Republic of Germany.

Table 1. Miscibility detected by DSC, molecular weights (GPC), and glass temperatures (T_g) of the polymers.

Polymer	M_n g/mol	M_w g/mol	T_g °C	Miscibility with PVC*
PVC	15900	40700	88	
PMMA	19600	35800	120	+
PEMA	76600	94200	75	+
PnBMA	39200	64300	31	+
PtBMA	118300	158800	116	−
PHMA	34700	63900	−7	+/−
PDMA	31300	79200	−59	−

* +: miscible; −: immiscible; +/−: miscible not over the whole range of concentration.

For mixing, the polymers were dissolved in 1,2-dichloroethane in 1 wt.% concentration. To obtain the given polymer ratios, adequate amounts of the solutions were combined, stirred for some minutes and then sprayed into a fifteen fold excess of methanol which was vigorously stirred. Figure 1 shows the apparatus, which was constructed following a proposal of Schulz and Sabel [11]. The advantage of this method is the fast and quantitative removal of the solvent. The samples can be dried in vacuo at room temperature for 24 hr and no remaining solvent is detected by thermogravimetry. The coarse-disperse samples were prepared by spraying the polymer solutions one after the other into the same beaker of methanol.

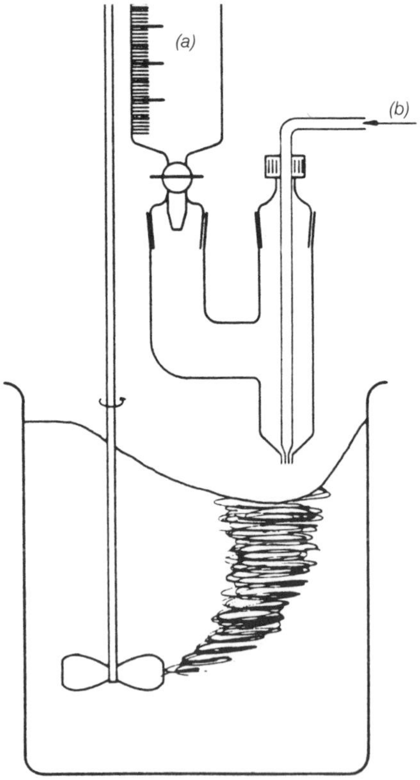

The thermal stability was measured on pressed disks (prepared at room temperature in equipment usually used for KBr pellets) with about 100 mg of the samples at 180°C in a steady flow of dry and oxygen free nitrogen. The amount of HCl was measured by conductometry. The method has been developed by Braun and Thallmaier [12]; details have been given more recently [13].

In some experiments the amount of HCl detected by conductometry was checked by titrating the HCl according to the Volhard method.

For recording UV-Vis spectra of degraded polymers, 100 mg of sample were degraded for 30 min. Afterward, the samples were dissolved in 10 ml of freshly distilled, peroxide free THF. This preparation was done in an argon atmosphere. The spectra were recorded in cells of 1 cm path length with a Perkin Elmer 554 UV-Vis spectrometer. The scanning speed was 60 nm/min.

The stabilised samples were prepared by weighing the amounts of mixture and Irgastab 17 M (dibutyltin-bis-isooctylthioglycollate, kindly supplied by Ciba Geigy, Marienberg) and putting them into an agate ball mill. The samples were milled for 1 hr. The degradation experiments were carried out as above.

(a)

(b)

Figure 1. Apparatus for preparing the mixtures: (a) polymer solution; (b) pressurised air.

RESULTS AND DISCUSSION

Thermal Stability

The thermal stability of the blends was measured at a constant temperature (180°C) in an inert atmosphere. The inert atmosphere was used to exclude disturbing reactions of oxygen. Working at a constant temperature is more related to the real conditions of processing than thermogravimetry, which is also used in literature [10]. Thermogravimetry is usually conducted at a certain heating rate and the temperature of the fast and nearly complete HCl loss of PVC is measured. This loss usually occurs at about 280°C. The processing temperatures for PVC are normally centered around 180°C, so that thermogravimetric results are not very useful for processing.

Recently, the question was raised [10] whether the thermal stability of PVC in mixtures is influenced by the miscibility of the system. Therefore, the miscibility of the systems was detected by glass transition (T_g) measurements using DSC. The appearance of one single T_g was taken as the evidence for miscibility [14]. The results are given in Table 1 together with the molecular weights (GPC) and T_g of the polymers. Full details will be given later [15]. To check the compatibility of the polymers at the temperature of degradation, the phase diagrams of the systems used in this investigation have been measured. Figure 2 shows these phase diagrams. The princi-

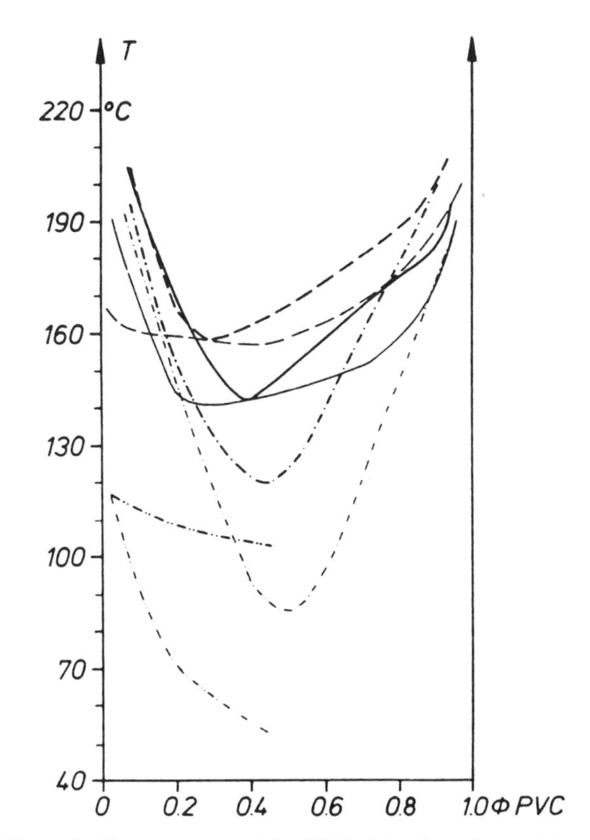

Figure 2. Phase diagrams of the PVC/poly(methacrylate) mixtures. ——— (binodal), ——— (spinodal) PMMA; ---------, --------- PEMA; -·-·-·-·-, -·-·-·-·- PnBMA; ··-··-··-, ··-··-··- PHMA.

ples of the measurement have been described by Wendorff et al. [16]; details of the measurements will be published soon [15].

According to Table 1 and Figure 2, PDMA and PtBMA are immiscible with PVC, PHMA is miscible only up to 50 wt.% of PVC, the other poly(methacrylates) are completely or at least partly miscible in the temperature range given in Figure 2. The phase diagrams indicate that most of the mixtures are partly demixed at the temperature of the degradation experiments (180°C).

Figure 3 shows the degradation curves of the PVC/PMMA mixtures. All samples are less stable than pure PVC. In contrast, PVC/PMMA samples, which are not homogeneously mixed, are more stable than PVC. The stability increases with decreasing amount of PVC in the mixture. A linear correlation between the dehydrochlorination after 1 hr and the content of PMMA in the coarse-disperse samples (later indicated as unmixed) appears in Figure 4.

When a disc of PVC and one of PMMA (both weighing about 100 mg) are brought in good contact and degraded in the usual manner, there is no change in the degradation behaviour of PVC. An explanation for this attitude of unmixed samples may be that the HCl is captured in the dispersed PMMA-particles. Obviously, during degradation of the unmixed samples, mixing must be in agreement with the phase diagram. This process of mixing is controlled by diffusion and is therefore relatively slow. It seems to have no influence on the degradation process of PVC.

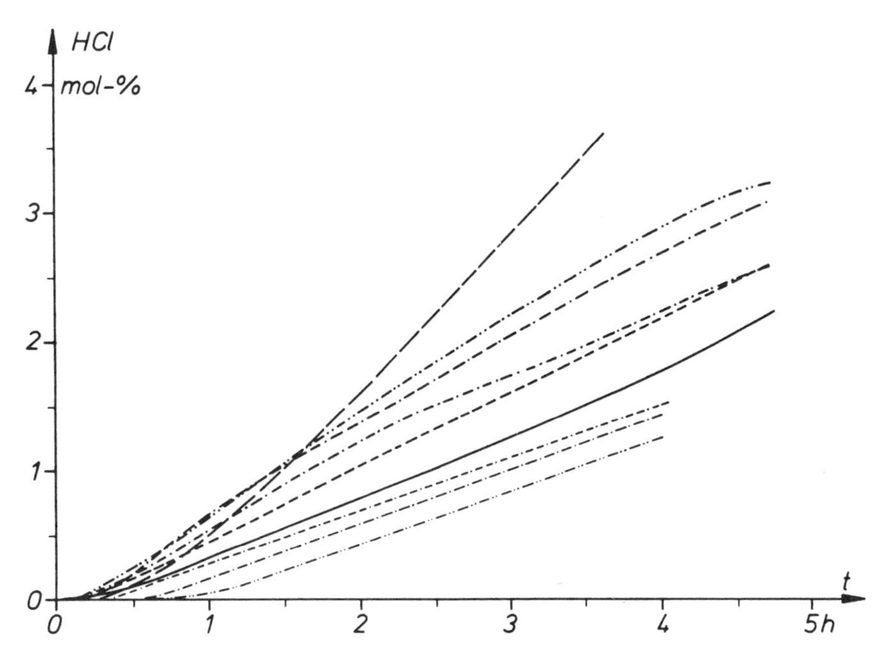

Figure 3. Time conversion curves for the dehydrochlorination of PVC/PMMA (wt./wt.) mixtures at 180°C in nitrogen.
——— PVC, --------- 90/10, -·-·-·- 75/25, -·-·-·-·- 50/50, -··-··-·· 25/75, — — 10/90.
Coarse-disperse samples: -·-·-·-·- 75/25, -·-·-·-·- 50/50,-··-··-··- 25/75.

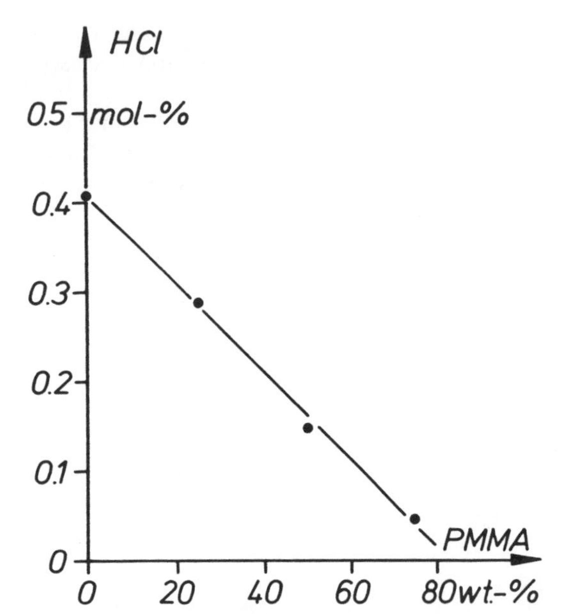

Figure 4. Correlation between the dehydrochlorination of PVC after 1 hr (mol-%) and the content of PMMA (wt.%) in the coarse-disperse PVC/PMMA samples.

The time conversion curves for the thermal degradation of the PVC/PEMA mixtures are shown in Figure 5. First, the stability decreases with decreasing PVC content. With 50 and less wt.% PVC the stability increases again, so that the blend with the highest PEMA content is the most stable one. According to Figure 6 the system PVC/PnBMA exhibits the same trend, but in this case the blends with 75 and 90 wt.% PnBMA are more stable than PVC. The mixtures with PHMA and PDMA behave in the same way. Therefore, especially high concentrations of poly(methacrylate)s in PVC may act as stabilisers toward the dehydrochlorination of PVC.

PtBMA was chosen as the second component in PVC mixtures to investigate the influence of branches in the ester side chains. The most obvious differences to PnBMA are the higher T_g and the immiscibility with PVC. These differences show that the influence of the side groups of polymers on the mixing behaviour is very strong. Figure 7 gives the time conversion curves for the dehydrochlorination of the PVC/PtBMA mixtures. According to this figure, the PVC/PtBMA blends behave just opposite to the mixtures with n-alkyl ester chains. The degradation is faster the higher the content of PtBMA is, so PtBMA exhibits antistabilising properties towards PVC.

To ensure that no products other than HCl (which may cause conductivity) are evolved, the HCl absorbed in water was titrated (according to the method of Volhard). This titration was done for the PVC/PtBMA-50/50 mixture after 5 hr of degradation; the difference between the conductometric and the titrimetric method was about 2.5%. Figuring the fact that the total amount of HCl is very low (about 3 mg), the error in the titration method is rather high. So it can be concluded that no significant

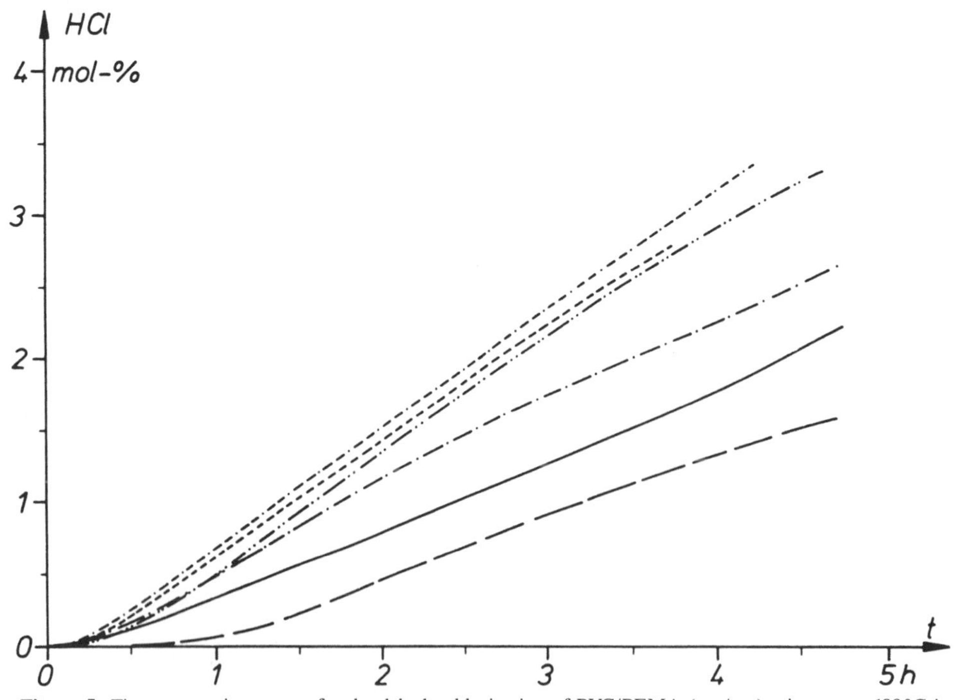

Figure 5. Time conversion curves for the dehydrochlorination of PVC/PEMA (wt./wt.) mixtures at 180°C in nitrogen.
——— PVC, - - - - - - 90/10, —·—·—· 75/25, -·-·-·-· 50/50, —··—··— 25/75, —— —— 10/90.

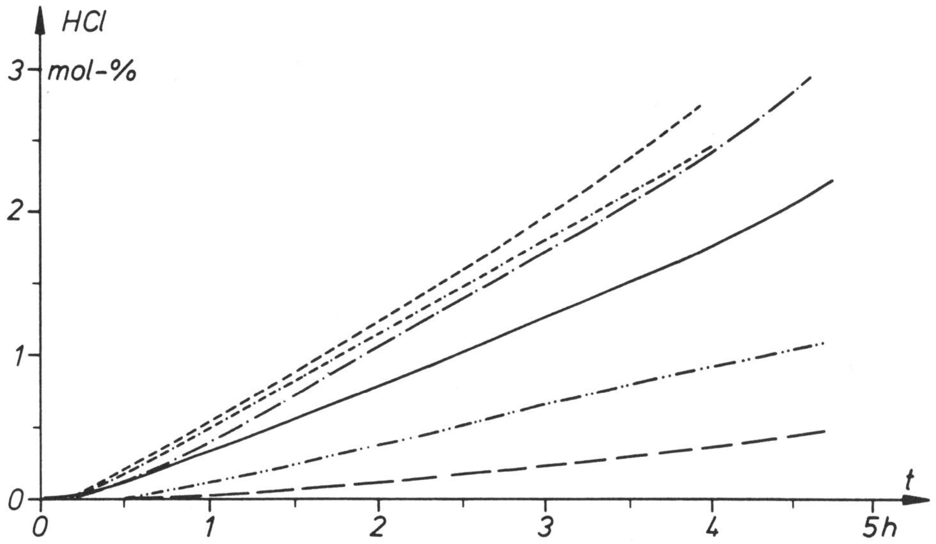

Figure 6. Time conversion curves for the dehydrochlorination of PVC/PnBMA (wt./wt.) mixtures at 180°C in nitrogen.
——— PVC, --------- 90/10, —·—·—·— 75/25, —·—·—·— 50/50, —··—··— 25/75, — — 10/90.

amounts of products which cause conductivity in water other than HCl are split off.

The question arises whether there is some kind of correlation between the thermal stability of the PVC/poly(n-alkyl-methacrylate) mixtures and the type of poly(methacrylate) used. To identify the poly(methacrylate)s, the number of carbon atoms in the ester chain can be used. To describe the degradation the yield of HCl (mol%) after 1 hr at 180°C was used.

For the samples with 50 and more wt.% PVC no trend

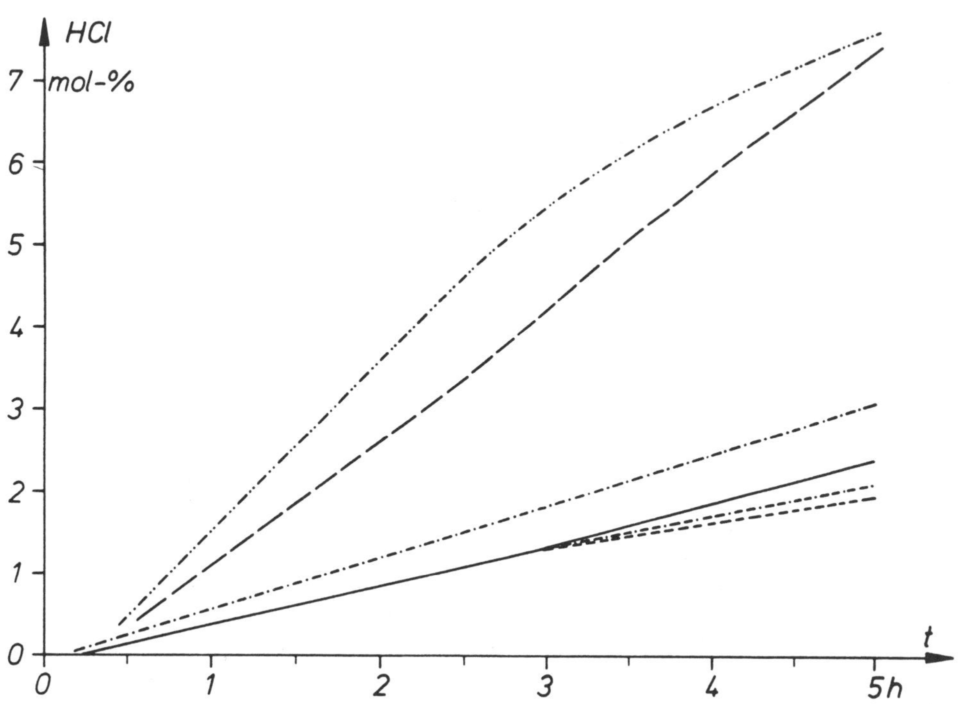

Figure 7. Time conversion curves for the dehydrochlorination of PVC/PtBMA (wt./wt.) mixtures at 180°C in nitrogen.
——— PVC, --------- 90/10, —·—·—·— 75/25, —·—·—·— 50/50, —··—··— 25/75, ——— 10/90.

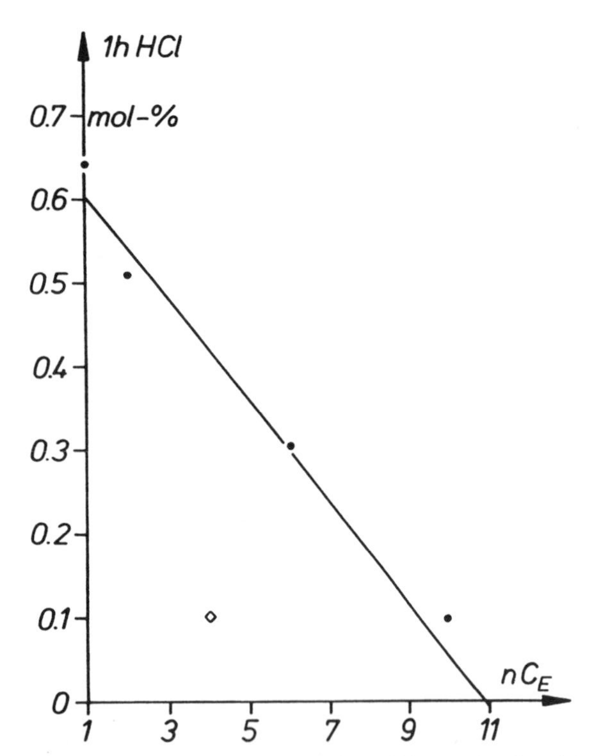

Figure 8. HCl split off after 1 hr (mol%) at 180°C as a function of the number of carbon atoms in the n-alkyl ester groups (nC_E) of PVC/poly(methacrylate)-25/75 mixtures. The value for PnBMA was omitted when drawing the line.

can be seen; this fact will be discussed later on. The behaviour of the mixtures with a high content of poly(n-alkyl-methacrylate) may be explained more easily. From Figures 3, 5 and 6 a stabilising effect occurring with longer ester side chains can be derived. The HCl split off after 1 hr at 180°C as a function of the number of carbon atoms in the ester group of the PVC/poly(n-alkylmethacrylate)-25/75 mixtures is given in Figure 8. If PnBMA is not taken into account, there is a linear dependence between conversion and length of the ester. The degradation drops with longer ester groups. The blend with PnBMA exhibits a very low conversion after 1 hr. This effect is in agreement with Figure 6, which shows the stabilising effect of PnBMA in PVC at high concentrations of PnBMA. The effect is more pronounced for PnBMA than for the other polymers.

UV-Vis-Spectroscopy of Degraded Samples

It is well-known that the dehydrochlorination of PVC proceeds via a zip-like mechanism, so that sequences of conjugated polyenes arise [1–3].

The UV-Vis spectra of such polyenes have been investigated for a long time beginning with Naylor and Whiting [17]. Generally speaking, the longer the polyene, the longer is the wavelength of the maximum absorption.

Table 2. Wavelengths of maximum absorption in UV-Vis spectra of degraded PVC/polymethacrylate blends (30 min, 180°C, N_2) in THF. Two values are given where an additional absorption below 370 nm appears; the first number indicates the strongest absorption.

PVC/ wt./wt.	PMMA	PEMA	PnBMA	PHMA	PDMA	PtBMA
	wavelength of maximum absorption (nm)					
10/90	460	464	365/464	305/410	304/464	435
25/75	456	455	463/365	330/412	364/460	435
50/50	435	455	435	303/384	455	435
75/25	435	455	435	435	435	435
90/10	435	435	434	435	410	435
PVC			387			

Polyenes of different lengths exist in degraded PVC, so that the UV-Vis spectra of degraded PVC are rather complicated. Braun and Sonderhof [18] analysed this problem and found that a quantitative interpretation of the spectra is very difficult due to the superposition of different absorption maxima. Therefore only qualitative explanations will be given in this work.

In all blends of PVC with poly(methacrylate)s the maximum absorption is shifted to longer wavelengths. But for the mixtures where a stabilising effect on the HCl elimination has been found above, an absorption below 370 nm appears (Table 2). PtBMA exhibits only a destabilising effect, and will not be taken into account further on.

Figure 9 shows the UV-Vis spectra of some PVC/PMMA-blends. It can clearly be seen how the absorption maximum shifts to higher wavelengths, and that the absorption at this maximum increases compared to pure PVC. This means the polyenes formed in the PVC in such mixtures are longer than in pure PVC, corresponding to a destabilising effect. Only with PMMA and PtBMA the effect is so clear and may be seen from Figures 1, 3 and 7 as well.

The spectra of the PVC/PHMA mixtures are given in Figure 10. For high concentrations of PVC, the absorption maximum shifts to longer wavelengths. With decreasing PVC content, the absorption at smaller wavelengths increases and overcomes the absorption at longer wavelengths. Thus, at higher concentrations PHMA (and likewise PnBMA and PDMA) have stabilising properties onto PVC. One effect of a stabiliser for PVC is to stop the growth of the polyenes when they are still small, as is the case with polymers with longer alkylester groups.

Figure 11 shows the trend in the poly(methacrylate) series for the 50/50 wt./wt. mixtures. For PMMA and PEMA a destabilising trend is visible, which is smaller for PEMA, meaning that the absorption at the maximum is smaller than with PMMA. For the mixtures with PnBMA, PHMA, and PDMA a certain stabilising effect is visible. The absorption maximum is decreasing and shifting to shorter wavelengths. In this series PHMA ex-

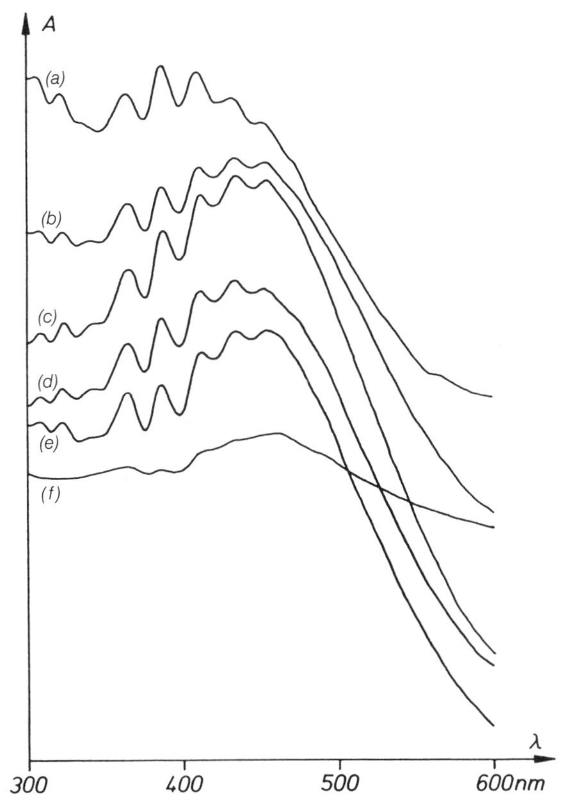

Figure 9. UV-Vis spectra of degraded PVC/PMMA (wt./wt.)-blends (30 min, 180°C, N_2, 10 g/l); spectra were recorded in THF solution. (a) PVC; (b) 90/10; (c) 75/25; (d) 50/50; (e) 25/75; (f) 10/90.

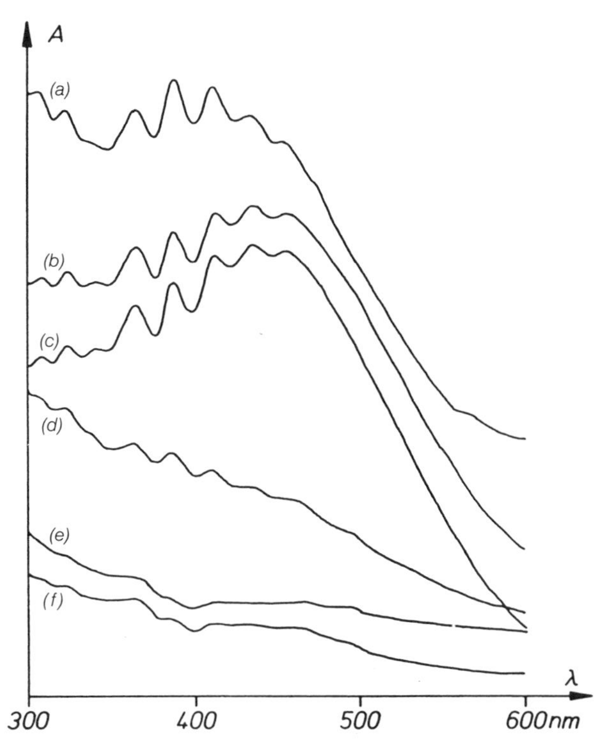

Figure 10. UV-Vis spectra of degraded PVC/PHMA (wt./wt.)-blends (30 min, 180°C, N_2, 10 g/l); spectra were recorded in THF solution. (a) PVC; (b) 90/10; (c) 75/25; (d) 50/50; (e) 25/75; (f) 10/90.

hibits the strongest effect; the absorption decreases from short wavelengths to longer ones, corresponding to the length of polyenes.

Figure 11 therefore illustrates the influence of the ester group on the thermal stability of PVC in blends. For short chain poly(methacrylate)s a destabilising effect occurs, whereas longer ester groups act as stabilisers.

In Figure 4 the correlation between the thermal stability and the number of carbon atoms in the ester group for the 25/75 mixtures is shown. For the samples with higher PVC contents no clear correlation was obtained. The UV-Vis spectra give a simple explanation therefore. At higher concentrations of PVC the destabilising effect of the poly(methacrylate)s dominates. This effect seems to be rather equal for all polymers. But the stabilising effect is dependent on the length of the ester group. So the cooperation of both effects leads to a complex behaviour. Only at high concentrations of the poly(methacrylate)s, where the stabilising effect dominates, is a clear correlation obtained.

One may discuss also an influence of the miscibility of the polymers, which decreases with longer ester chains. But according to the phase diagrams in Figure 2, at the temperature of degradation (180°C) nearly all of the mixtures are demixed or partly demixed. Because of the

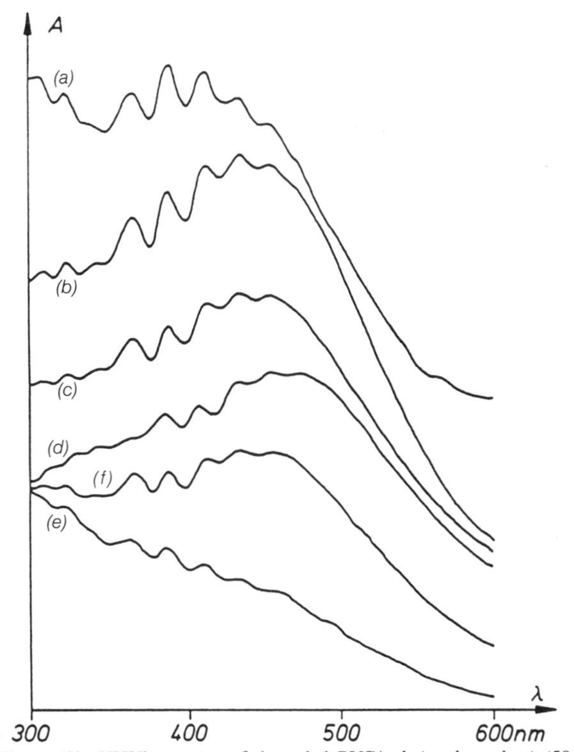

Figure 11. UV-Vis spectra of degraded PVC/poly(methacrylate)-(50/50)-blends (30 min, 180°C, N_2, 10 g/l); spectra were recorded in THF solution. (a) PVC; (b) PVC/PMMA; (c) PVC/PEMA; (d) PVC/Pn-BMA; (e) PVC/PHMA; (f) PVC/PDMA.

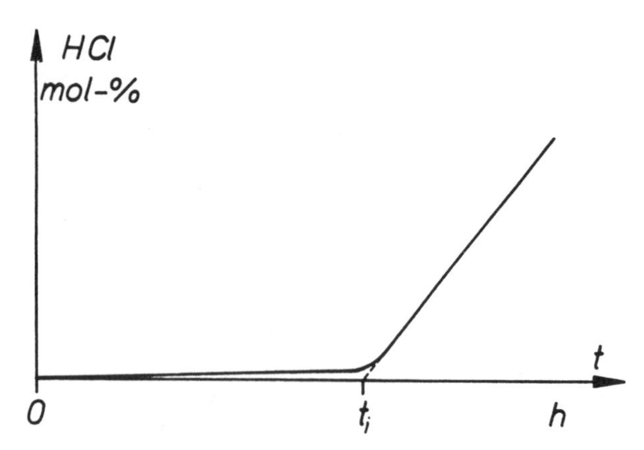

Figure 12. Principle dehydrochlorination curve of a stabilised PVC and calculation of the induction time (t_i).

shape of the phase diagrams the effect of the miscibility should be somewhat symmetric to the composition of the mixtures. Therefore, an influence of the miscibility on the thermal stability of the PVC is very unlikely.

Stabilisation Experiments

Some stabilisation experiments have been carried out with PVC/PnBMA and PVC/PMMA (mixed and unmixed). The stabiliser used was Irgastab 17 M (dibutyltin-bis(iso-octylthioglycollate)). The advantage of this stabiliser is that it needs no costabiliser to obtain good results and that it is usable for all purposes. The amount of stabiliser was about 1 wt.% referred to PVC. All values given in this work are corrected to 1 wt.% stabiliser.

The dehydrochlorination curve of stabilised PVC shows an induction period where almost no HCl is evolved (Figure 12). During this period the stabiliser is consumed. When nearly all of the stabiliser has reacted, the sample starts to split off HCl. The HCl evolution may then be slower, faster or the same as for pure PVC – due to the reaction products of the stabiliser. In the case of Irgastab 17 M the dehydrochlorination after the induction period is a little bit slower than the dehydrochlorination of unstabilised PVC.

Figure 12 shows schematically the dehydrochlorination of stabilised PVC. In this figure it is also demonstrated how the induction time was calculated.

Table 3 gives the induction times for the PVC/PMMA-system. Corresponding to Figure 2, the unmixed samples exhibit induction periods even without stabiliser (t_{i0} in Table 3). When the values without stabiliser (t_{i0}) are subtracted from the values for stabilised samples (t_i), the same value as for PVC is obtained for all unmixed samples within the range of error (last row in Table 3). That means that the two stabilising effects are additive. The absorption of HCl in the PMMA particles and the effect of the stabiliser are separated from each other. To get the effect this clearly, the stabiliser must be assumed to be located, nearly quantitatively, in the PVC particles.

The homogeneously mixed samples differ in their behaviour from the unmixed samples. All mixtures exhibit a longer induction period than PVC. Therefore a synergistic effect between Irgastab 17 M and PMMA has to be assumed. A linear correlation between t_i and the PMMA content appears in Figure 13, but pure, stabilised PVC does not participate in this correlation. The smaller the content of PMMA in the mixture, the more stable is the PVC.

There is a simple explanation for this finding. The amount of stabiliser is referred to the amount of PVC, and the ratio stabiliser/PMMA decreases with increasing amount of PMMA. PMMA alone destabilises PVC (Figure 3). So two counter-current effects must be taken into account: PMMA on one hand improves the properties of the stabiliser and on the other hand destabilises PVC. With increasing amounts of PMMA the destabilising properties increase, whereas the influence of the stabiliser decreases because the ratio stabiliser/PMMA decreases. So the behaviour may be easily explained in this way.

As the curves of dehydrochlorination (Figures 3, 5, and 6) show, high amounts of PnBMA exhibit the best stabi-

Table 3. Induction times for the system PVC/PMMA. For the coarse-disperse samples t_{i0} means the induction period of the samples without stabiliser. t_i is the induction time of samples stabilised with 1 wt.% Irgastab 17 M.

PVC/PMMA wt./wt.	Mixed t_i min	Unmixed t_{i0} min	Unmixed t_i min	Unmixed $t_i - t_{i0}$ min
25/75	65	50	111	61
50/50	88	24	84	60
75/25	128	7	71	64
PVC	61	0	61	61

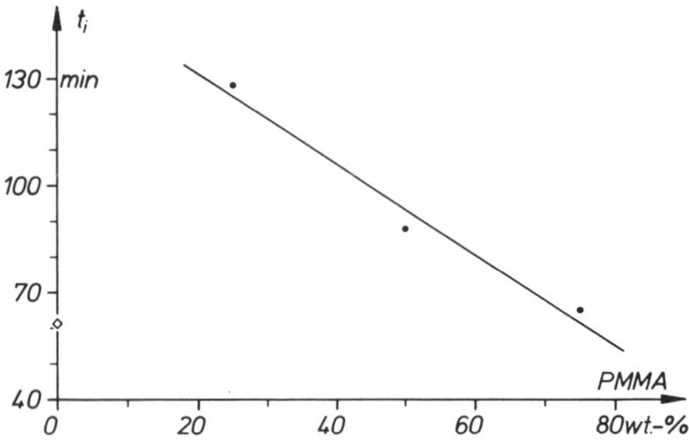

Figure 13. Induction time (t_i) as a function of the PMMA content in stabilised PVC/PMMA blends at 180°C.

Table 4. Induction times (t_i) of the PVC/PnBMA mixtures stabilised with 1 wt.% Irgastab 17 M, degraded at 180°C in nitrogen.

PVC/PnBMA wt./wt.	t_i min
25/75	not detectable
50/50	166
75/25	260
PVC	61

lising properties of all poly(methacrylate)s. Therefore, stabilising experiments with this polymer are interesting. The procedures were the same as with the PVC/PMMA-blends. The induction times are summarized in Table 4.

Unfortunately, an induction time for the mixture with 75 wt.% PnBMA could not be found. The change in the slope in the region of t_i is so small that it cannot be detected. This lack of change is due to the stabilising effect of PnBMA (Figure 6). According to Table 4, PnBMA exhibits a synergistic behaviour with the tin stabiliser. Small amounts of PnBMA have the best costabilising properties. In both cases, the behaviour is the same as for the PMMA mixtures, but PnBMA is far more effective than PMMA.

CONCLUSIONS

As the dehydrochlorination curves and the UV-Vis spectra show, all poly(methacrylate)s used have an influence on the degradation of PVC in blends. Longer n-alkylester groups and higher concentrations of the corresponding poly(methacrylate) exhibit some stabilisation of PVC, whereas smaller ester chains and low concentrations yield some destabilisation. Both effects compete so that no clear trend throughout the series of poly(methacrylate)s is obtained.

An effect of the miscibility of the samples on the thermal stability cannot be detected, in contradiction to Goulet and Prud'Homme [10]. But they used different polymers in mixtures with PVC and thermogravimetry for detecting the thermal stability. Additionally, they did not know the phase diagrams of their blend systems. So their results and this work may not be compared directly.

The stabilising experiments with dibutyltin-bis(isooctylthioglycollate) show small amounts of poly(methacrylate)s to act as costabilisers and to improve the thermal stability significantly. Therefore poly(methacrylate)s, which are used as processing aids for rigid PVC, may have a second favourable property—together with special types of stabilisers they improve the thermal stability of PVC.

ACKNOWLEDGEMENTS

Financial support by the Bundesminister für Wirtschaft through the Arbeitsgemeinschaft Industrieller Forschungsvereinigungen e. V. (AIF) is gratefully acknowledged.

REFERENCES

1. Braun, D. and E. Bezdadea. "Theory of Degradation and Stabilisation Mechanisms," in *Encyclopedia of PVC*. 2nd ed. L. I. Nass, ed. New York:Dekker, p. 397 (1986).
2. Starnes, W. H., Jr. "Mechanistic Aspects of the Degradation and Stabilisation of Poly(Vinyl Chloride)," in *Developments in Polymer Degradation, Vol. 3*. N. Grassie, ed. London:Applied Science Publishers, p. 135 (1981).
3. Hjertberg, T. and E. M. Sörvik. "Thermal Degradation of PVC," in *Degradation and Stabilisation of PVC*. E. D. Owen, ed. London:Elsevier, p. 21 (1984).
4. Paul, D. R. and J. W. Barlow. *J. Macromol. Sci.—Rev. Macromol. Chem.*, C18:109 (1980).
5. Walsh, D. J. and J. G. McKeown. *Polymer*, 21:1330, 1335 (1980).
6. Walsh, D. J. and G. L. Cheng. *Polymer*, 25:499 (1984).
7. Tremblay, C. and R. E. Prud'Homme. *J. Polym. Sci., Polym. Phys. Ed.*, 22:1857 (1984).
8. Lutz, J. T., Jr. "Polymeric Additives for Poly(Vinyl Chloride)," in *Degradation and Stabilisation of PVC*. E. D. Owen, ed. London:Elsevier, p. 253 (1984).
9. McNeill, I. C. and D. Neil. *Eur. Polym. J.*, 6:143, 569 (1970).
10. Goulet, L. and R. E. Prud'Homme. *Eur. Polym. J.*, 22: 529 (1986).
11. Schulz, R. C. and A. Sabel. *Makromol. Chem.*, 14:115 (1954).
12. Braun, D. and M. Thallmaier. *Kunststoffe*, 56:80 (1966).
13. Braun, D., B. Böhringer, B. Iván, T. Kelen and F. Tüdos. *Eur. Polym. J.*, 22:1 (1986).
14. Olabisi, O., L. M. Robeson and M. T. Shaw. *Polymer–Polymer Miscibility*. New York:Academic Press, p. 133 (1979).
15. Braun, D., B. Böhringer, W. Knoll and N. Eidam. To be published.
16. Ebert, M., R. W. Garbella and J. H. Wendorff. *Makromol. Chem., Rapid Commun.*, 7:65 (1986).
17. Naylor, P. and M. C. Whiting. *J. Chem. Soc.*, 3037 (1955).
18. Braun, D. and D. Sonderhof. *Polym. Bull.*, 14:39 (1985).

C. CROUZET[1]
S. ZEHNACKER[1]
J. MARCHAL[1]

About Antioxidant Activity of Hindered Amine Additives to Understand Polyolefin Stabilization

ABSTRACT

The kinetics of the initial period of the γ-radiation-induced oxidation of well-oxygenated diluted solutions of 2,2,6,6-tetramethylpiperidine (TMPH) or Tinuvin 770 in 2,4-dimethylpentane (DMP), in the dark at 25°C, showed that secondary hindered amine stabilizers (sec.HAS) are not antioxidants (AO), because they do not stop the propagation of oxidative chain reactions. Their participation in the propagation reactions includes the production of nitroxyl radicals (>NO·). They also showed that sec.HAS are not hydroperoxide decomposers.

The kinetics of the initial period of the radiation-induced oxidation of well-oxygenated diluted solutions of the nitroxyl derivatives (TMPO· and ·ON--NO·) of these sec.HAS and of Ionol (BHT) phenolic AO in DMP were quite similar. These experiments showed that >NO· radicals have strong AO properties which cannot be explained by alkyl radical or H· atom scavenging, because the oxygenation of the solutions was insured.

The kinetics of the thermo-oxidation, in a forced air oven at 95°C, of thin blown films of LDPE + Tinuvin 770 were followed during 254 days. At the beginning (50 days), the >NO· concentration increased without induction period, reached a maximum and decreased to very low values: this kinetic behavior is in agreement with the above conclusion concerning AO properties of >NO· radicals. Over all the experiments, the rate of the accumulation of carbonyl groups measured at ν_{max} = 1718 cm^{-1} remained low and nearly constant. At the very beginning, the results were perturbed by the evaporation of the largest amount of this additive segregated in microphases where ν_{max} = 1718 cm^{-1} corresponds to >C=O stretching of its ester groups. Over all the experiments, the concentrations of hydroperoxides and other functional groups absorbing in the O–H stretching region of the IR spectrum remained too small to be measurable.

Antagonism was found between this sec.HAS and Irganox 1076 phenolic AO at the concentration usually used for polymer processing. This interaction can be explained by already known mechanisms.

Thermo-oxidation of thin blown films of LDPE + Tinuvin 622 was studied in the same conditions during 107 days. It appeared that tertiary HAS are very efficient AO since no chemical change was observed either by IR spectrophotometry or by EPR spectrometry, till the end of the heating treatment. No antagonism was observed between this tert.HAS and Irganox 1076 in experiments pursued up to 254 days. This result could be a strong argument against the production of >NO· radicals, at least by a primary reaction, in the reaction scheme of the AO stabilization of polyolefins and other organic compounds by tert.HAS, which remains unsolved.

KEY WORDS

Polyolefin stabilization, antioxidants, mechanisms, hindered amine stabilizers, nitroxyl radicals, phenolic antioxidants, 2,4-dimethylpentane, LDPE, radiation-induced oxidation, thermo-oxidation.

1. INTRODUCTION

The secondary or tertiary hindered amine stabilizers (HAS) derived from 2,2,6,6-tetramethylpiperidine are substances which are used as additives in polyolefines and give marked stabilization to natural aging: solar light (λ > 290–300 nm) with consequent heating. . . . The properties imparted by these HAS have often been interpreted in terms of photostabilizing (HALS) concepts. The HAS have no intrinsic photophysical or photochemical properties which could be directly responsible for their stabilizing behavior [1]. HAS are, however, well-known as antioxidants (AO) [2,3].

To pursue the classification of these additives among photostabilizers (although they are antioxidants with numerous properties) is no longer justified. Even though it is perhaps a question of terminology, it is nonetheless unfortunate. The confusion that this has entailed is probably responsible for the slow progress in the understanding of stabilizer reaction mechanisms.

The major drawback of studying these HAS in a photoinitiated process is that the photochemical behavior of the reaction intermediate and degradation products com-

[1]Laboratoire d'Etude de la Degradation et de la Stabilisation des Polymeres. Institut Charles Sadron (CRM-EAHP), CNRS-ULP 6 rue Boussingault, 67083 Strasbourg, Cedex, France.

plicate the reaction schemes of the studied compounds. It thus becomes very hard, if not impossible, to find a clear interpretation of the experimental results.

For this reason, it is more logical to study the stabilization mechanisms of organic materials by this AO class by using initiation processes of oxidative degradation of these materials, excluding any photochemical effect. This is why the experiments performed in our laboratory were done in the dark, either by radiation-induced oxidation at 25°C or by thermally induced oxidation.

Radiation-induced initiation by exposure to high-energy electromagnetic radiation has numerous advantages in the study of oxidative degradation mechanisms of a given substrate. Therefore effects due to additives can easily be studied. It can also be used at any temperature including room temperature at which secondary reactions leading to thermal decomposition of the oxidative degradation products are definitely negligible. Being nonselective, this initiation process makes interpretation of the results in terms of chemical processes considerably easier for the following reasons. First of all, one can neglect primary reactions involving parasitic functional groups of the substrate or the additives. Also, secondary reactions involving primary products of the oxidative degradation of the substrate or of the additives can be neglected during the initial period. These simplifications, which are very useful in determining the chemical processes, are not at all possible when photochemical initiation is used.

Another advantage of Co60 γ-ray initiation is that the kinetic studies are easier due to the quite weak absorption of these photons. Thus, it is not necessary to account for profiles of absorption with sample thickness. At present, the only restriction is to work with a sufficiently small absorbed dose rate in order to ensure that the consumption of oxygen by the oxidative reactions is not faster than diffusion of this gas from the environment.

2. EXPERIMENTAL

2.1. Products

- LDPE high pressure (radical polymerization) in pellets (*MI* = 2) without additive, provided by Charbonnages de France Chimie
- LDPE high pressure, in films (*MI* = 0.3 or 0.2) with or without additives, provided by Atochem
- 2,4-dimethylpentane (DMP), 99.5% (Baker), which has been purified on a column of silica gel/aluminum-oxide W 200 basic (Woehlm) followed by Gas-chromatography (GC) analysis
- triphenylphosphine ($\varnothing_3P$) (Fluka)
- 2,4-dimethyl 2-pentanol (HO–R–H) (Fluka)
- additives:

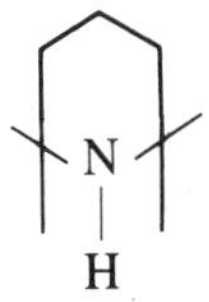

N–H

(T M P H)

2,2,6,6-tetramethyl-piperidine
(Fluka)

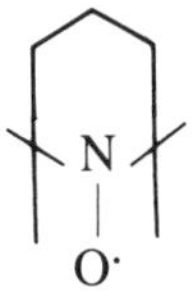

N–O·

(T M P O·)

2,2,6,6-tetramethyl-1-oxylpiperidine
(Aldrich)

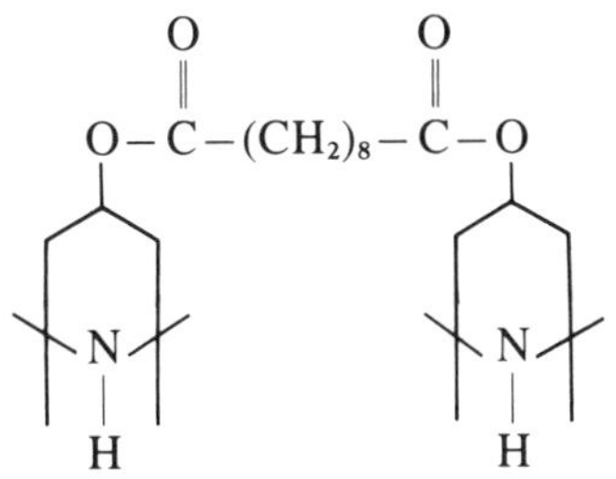

$O-C(=O)-(CH_2)_8-C(=O)-O$

N–H N–H

Tinuvin 770 HN=NH)
(Ciba Geigy)

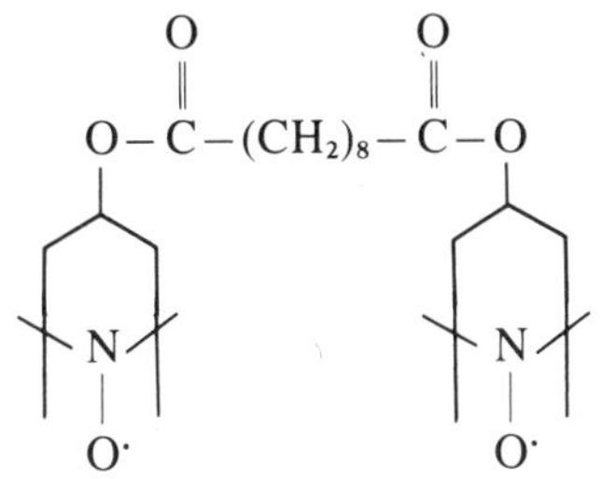

$O-C(=O)-(CH_2)_8-C(=O)-O$

N–O· N–O·

(·ON--NO·)
(Chemopetrol)

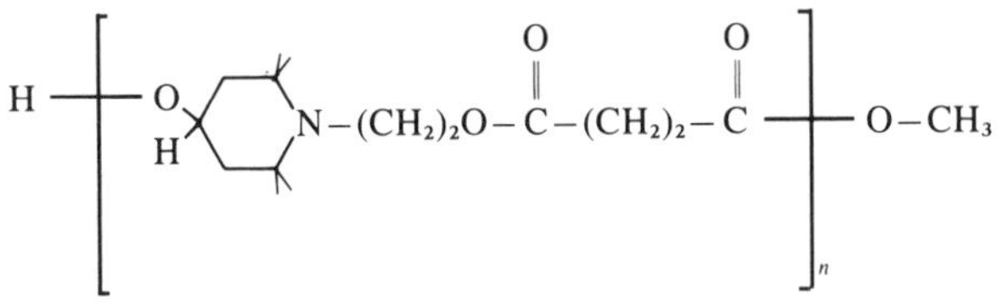

$H-[O-\ldots N-(CH_2)_2O-C(=O)-(CH_2)_2-C(=O)]_n-O-CH_3$

Tinuvin 622
(Ciba Geigy)

OH

Me

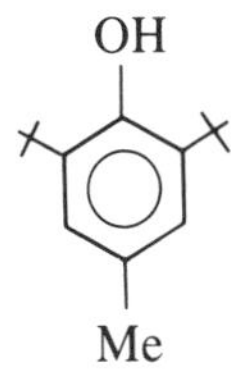

Ionol or BHT
(Chemopetrol)

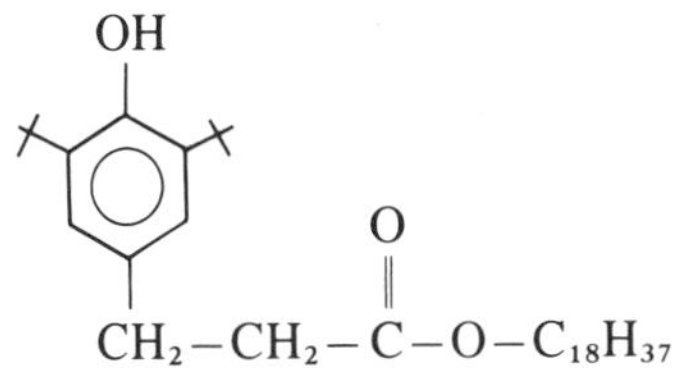

OH

$CH_2-CH_2-C(=O)-O-C_{18}H_{37}$

Irganox 1076
(Ciba Geigy)

2.2. Initiation Processes

- Radiation-induced oxidation of 2,4-dimethylpentane at 25°C.
  The experiments were done by exposure to Co60 γ-radiation from a panoramic source, absorbed dose rate I = 72 rad min.$^{-1}$, in open glass test tubes (external diameter 20 mm). The height of the liquid ($\leq$1 cm) was chosen in order to avoid the limitation of oxygenation of the solutions during irradiation due to the speed of diffusion of oxygen through the air/liquid interface.
  The radiation dosimetry has been done with the Fricke dosimeter followed by required corrections to calculate I in DMP [4].
- Thermo-induced oxidation of LDPE films at 95°C.
  The experiments were done in forced hot air ovens (HORO ovens and modified laboratory ovens).

2.3. Processing of LDPE Films

The first series of blown films were industrially prepared. The second were prepared in our laboratory with a device especially conceived to ensure all physico-chemical parameters of the process.

- Industrial films (60 ± 10 μm).

  ① LDPE (MI = 0.2) without additive
  ② LDPE (MI = 0.3) + Irganox 1076 (0.01 wt.% = 1.9 × 10^{-4} mole kg^{-1})
  ③ LDPE (MI = 0.3) + Irganox 1076 (0.01 wt.%) + Tinuvin 622 (0.5 wt.% = 1.8 × 10^{-2} mole kg^{-1} piperidyl group)

- Laboratory films–The polymer used was the LDPE in pellets, specially prepared without any additive, provided by CdF Chimie. Film blowing was done at 190°C with a mini-press conceived and built in our laboratory with the ability to perform all the steps: melting, extrusion, and film-blowing, in the complete absence of oxygen [5]. To obtain a homogeneous concentration of the additives in the films, the pellets were first impregnated with the additives. This was done by mixing the pellets and the powdered additives in a Rotovapor for 68 hours at 97–98°C and under CO_2. The films thus prepared were:

  – with thickness = 50 ± 10 μm

  ①' LDPE without additive
  ④ LDPE + Tinuvin 770 (0.5 wt.% = 2 × 10^{-2} mole kg^{-1} piperidyl group)
  ⑤ LDPE + Tinuvin 770 (0.5 wt.%) + Irganox 1076 (0.01 wt.% = 1.9 × 10^{-4} mole kg^{-1} phenolic group)

  – with thickness = 100 ± 30 μm

  ⑥ LDPE + Tinuvin 622 (0.5 wt.% = 1.8 × 10^{-2} mole kg^{-1} piperidyl group)

Films prepared with this polymeric tert.HAS have numerous structural irregularities. Because these defects do not interfere with the proposed experiments, it was unnecessary to prevent their formation by modifying the processing conditions. (This has been solved later.)

It is interesting to note that, under our laboratory conditions, we did not run into the well-known problem of the required addition of a phenolic AO to make possible the processing of sec.HAS containing polyolefins. This problem is related to the fact that HAS are generally considered as bad AO in melted systems [2,6,7]. Nevertheless, it will be seen later (§ 3.2 and 5.2) that sec.HAS cannot act as AO before >NO· radicals have been produced at sufficient level. In fact, the LDPE blown films we prepared with or without additive show neither a trace of oxidative degradation nor the formation of macroscopic defects (fish eyes) which is known to be induced by such degradation [5].

2.4. Analysis

2.4.1. GC ANALYSIS OF THE ALCOHOLS AND THE HYDROPEROXIDES PRODUCED BY THE OXIDATIVE DEGRADATION OF DMP

It is not possible to make the quantitative analysis of hydroperoxides when HAS or nitroxyl radicals >NO· are present, using the iodometric or thiocyanate methods [8].

Therefore, to determine the amounts of mono- and di-hydroperoxides as well as those of mono- and di-alcohols produced during the oxidation of DMP, an experimental procedure has been perfected in our laboratory. This method is based on the GC analysis of the mono- and di-alcohols in the samples before and after the $\varnothing_3P$ reduction of hydroperoxides.

This reaction, at 25°C, which is not accompanied by the reduction of dialkylperoxides [9,10], thus leads to increasing amounts of mono- and di-alcohols equal to the corresponding mono- and di-hydroperoxides. Experiments were done to verify that the addition of sec.HAS or >NO· radicals does not perturb the results.

The analyses were done with an HP 5830A GC under the following conditions:

column	UCON POLAR HB 2000 20%, Chromosorb W 60–80, length 3 m
column temperature	140°C, isothermal or temperature-programmed
vector gas	N_2, flux 25 cm^3 min.$^{-1}$
detector	FID
calibration	2,4-dimethyl-2 pentanol (HO–R–H) as received, and 2,4-dimethyl-2,4-pentanediol (HO–R–OH) obtained by $\varnothing_3P$ reduction of 2,4-dimethyl-2,4-dihydroperoxypentane (HOO–R–OOH) after titration by thiocyanate method [11]

2.4.2. ESR SPECTROMETRY OF NITROXYL RADICALS >NO·

Analyses were done with a Brucker ER 4420 spectrometer to follow [>NO·] concentration during:

- the radiation-induced oxidation of the solutions: DMP + sec.HAS (TMPH or Tinuvin 770)
- the thermo-oxidation of samples of LDPE films with sec. or tert.HAS (Tinuvin 770 or Tinuvin 622)

2.4.3. IR SPECTROPHOTOMETRY

The analyses of the LDPE films were done with a Perkin Elmer 583 spectrometer. The carbonyl index at $\nu_{max} = 1718\ cm^{-1}$ was calculated using the formula:

$$>C{=}O \text{ index } = \left(\frac{1}{d} \log \frac{I_0}{I_T}\right) \times 100$$

where:

I_0 is the entering light intensity
I_T is the transmitted light intensity
d is the thickness of the sample (μm)

The analysis of Tinuvin 770 powder was done with a Nicolet 60SX FTIR-spectrometer equipped with a multi-reflection device.

2.5. Physical State of Tinuvin 770 and Tinuvin 622 in LDPE Films

In order to consider the concentration distribution of the additives in the polymeric matrix, it is useful to recall some elementary thermodynamic properties. Compatibility and solubility will be considered as synonymous here. If the solubility of an additive in a polymer increases with temperature, the effect of quenching (depending upon the cooling rate) is characterized by a larger concentration of the additive remaining dispersed in the polymer at room temperature and a lower amount of the additive which is segregated in microphases. But it is a non-stable system which returns to equilibrium on increasing the temperature or annealing.

This phenomenon can be easily followed by IR spectroscopy when the dispersion of the additive in the polymer matrix is the result of the destruction of the interactions between polar groups which contribute to the structure of this additive in the solid state at the temperature considered. This is precisely the case of Tinuvin 770 in polyolefins. But this approach has not been studied until now because most of the publications concern polypropylene (PP) [6,12–16] and HDPE [17] in which this additive is soluble at usual concentrations and very few concern LDPE [13,14].

2.5.1. TINUVIN 770

Table 1 shows the IR absorption bands of this sec.HAS in different solvents and polyolefins. It is seen, as is well-known, that this HAS has very low solubility in LDPE but much higher solubility in PP [18]. It should be noted that it would be possible to determine the saturation concentration above which phase separation would occur by looking at blends with various concentrations of Tinuvin 770. Phase separation occurs when the 1718 cm^{-1} peak characterizing Tinuvin 770 microphases appears beside the 1736 cm^{-1} peak which characterizes the molecular dispersion of this HAS in LDPE.

In the following, it will be shown that this sec.HAS is only active in the molecular dispersed state, since stabilization of the polymer against thermal degradation at 95°C has been achieved at concentration much lower than the initial concentration (see § 4.1.1).

It is also interesting to note that the IR spectra of films

Table 1. ν_{max} of IR absorption bands of HAS at ambient temperature [(e): ν_{max} of the shoulder].

	wt.%	>C=O Ester			>N—H			Piperidyl-
Tinuvin 770								
• powder		1717 cm^{-1}			1217 cm^{-1}		1238 cm^{-1}	1165 cm^{-1}
• in KBr*		1720 cm^{-1}			1220 cm^{-1}		1242 cm^{-1}	1180–1160 cm^{-1}
• in KBr	0.6	1718 cm^{-1}			1219 cm^{-1}		1242 cm^{-1}	1165 cm^{-1}
• in hexane*				1739 cm^{-1}		1236 cm^{-1}		1180–1160 cm^{-1}
• in hexane	0.5		(e) 1735 cm^{-1}	1740 cm^{-1}		overlapping		overlapping
• in DMP	0.5		(e) 1735 cm^{-1}	1740 cm^{-1}				
• in atactic PP film*				1739 cm^{-1}		1236 cm^{-1}		1180–1160 cm^{-1}
• in isotactic PP discs	2			1736 cm^{-1}		overlapping		overlapping
• in LDPE films	0.5	1718 cm^{-1}		(e) 1737 cm^{-1}	1220 cm^{-1}		1242 cm^{-1}	1168 cm^{-1}
• in LDPE discs	1.5–2	1718 cm^{-1} ↗		↙ 1736 cm^{-1}	1219 cm^{-1} ↗		1241 cm^{-1}	1167 cm^{-1}
Tinuvin 622								
• in LDPE films	0.5			1736 cm^{-1}				1160 cm^{-1}

*D. J. Carlsson, D. M. Wiles [12].

④ and ⑤ (LDPE + Tinuvin 770 ± Irganox 1076) did not change for blown films stored at ambient temperature in the dark over long periods of time. On the other hand, the IR spectra of thin discs ($\sim 100\ \mu m$) quenched from supersaturated solid solutions of Tinuvin 770 (see preparation below) showed that the concentration of the dispersed phase diminished rapidly. The change in concentration was already measurable after a few days and approached the saturated value at ambient temperature, as observed in the films. Here it is seen that the band at 1718 cm^{-1} corresponding to microphase-separated HAS increases while the band at 1736 cm^{-1} corresponding to the molecular dispersion decreases. At the same time, the intensity of the >N−H band at 1219 cm^{-1} increased steadily while another amine band at 1241 cm^{-1} stayed at a constant intensity. There was also no measurable change in the band of the 2,2,6,6-tetramethylpiperidyl group at 1167 cm^{-1} [12]. It is also noted that a band from the C−O− ester bond at 1164 cm^{-1} overlaps the previously mentioned band.

The above observations lead to the conclusion that the blown films used here did not show the quenching phenomenon. These experiments also allowed for the assignment of the IR absorption bands.

The quenching phenomenon was not seen in the IR spectra of isotactic PP discs containing 2 wt.% Tinuvin 770. This shows that, at this concentration, the HAS remains completely molecularly dispersed in the polymer at room temperature.

2.5.2. TINUVIN 622

Table 1 shows that this tert.HAS is apparently more soluble in LDPE at ambient temperature than the sec.HAS previously discussed: ν_{max} = 1736 cm^{-1} as in PP films or discs, and no change was seen in the IR spectra after storage of these films in the dark.

2.5.3. PREPARATION OF LDPE AND PP DISCS

The discs were made with a press specially designed in this laboratory to work under vacuum with control over the following parameters: shape, disc thickness, temperature, pressure on the molten material, molding time, and cooling rate.

With LDPE (MI = 2), the discs were made by molding at 190°C for one hour (1/2 hour under 1 atm., then 1/2 hour under 6 kg cm^{-2}).

The discs were made from samples of films without additive blown in the laboratory and Tinuvin 770 powder sandwiched between them. After releasing the molding pressure (or leaving this pressure in certain cases), the samples (under vacuum) were cooled either slowly (by leaving the press in air) or quickly (by circulating water through a cooling coil, which allows for complete cooling in several minutes). The discs were removed after the press cooled to ambient temperature.

With the isotactic PP powder (Mosten, Chemopetrol, Czechoslovakia), the discs were prepared as above except that they were molded at 240°C. The blown film without additive ($\sim 50\ \mu m$) was prepared under similar condition, but at 240°C, to those of the LDPE films, using the same mini-press extruder.

3. KINETICS OF THE RADIATION-INDUCED OXIDATION OF sec.HAS SOLUTION IN 2,4-DIMETHYLPENTANE AT 25°C

3.1. Radiation-Induced Oxidation of DMP at 25°C

The choice of this model compound of PP, which is liquid at the experimental temperature, fulfills the following requirements:

(a) A small number of possible reactions, permitting the quantitative analysis of their products needed for the elucidation of the corresponding reaction scheme
(b) The elimination of any complications or uncertainties stemming from the effect of the crystalline or amorphous nature of the solid phase on the mobility of the reactive species (including oxygen)
(c) Furnishing the appropriate conditions for studying the competition between intra- and intermolecular propagation reactions in the oxidative chain reaction scheme

3.1.1. REACTION SCHEME OF THE RADIATION-INDUCED AUTOXIDATION OF DMP AT 25°C

The thermal oxidation process of DMP is well-known [19]; its simplicity illustrates the interest in this model compound.

DMP degrades to produce four main products: 2,4-dimethyl-2-pentanol (HO−R−H); 2,4-dimethyl-2,4-pentanediol (HO−R−OH); 2,4-dimethyl-2-hydroperoxypentane (HOO−R−H) and 2,4-dimethyl-2,4-dihydroperoxypentane (HOO−R−OOH). Small quantities of diperoxides resulting from termination reactions are also produced. The very low yield of carbonyl compounds produced by thermo-oxidation of DMP at 100°C results from the fact that dismutation reaction of tertiary alcoxyl radicals is not possible. It also shows that β-scission of these radicals is unfavoured.

The radiation-induced autoxidation of DMP at 25°C was performed under the previously described conditions with a radiation absorbed dose rate I = 72 rad $min.^{-1}$. The flexibility to choose the I value and thus the autoxidation initiation rate (which is constant and depends uniquely upon the distance from the γ source) is the main advantage of a panoramic Co60 γ-radiation source. With this technique, it is possible to reduce the initiation rate to the domain where the autoxidation is not controlled by the rate of oxygen diffusion from the atmosphere to the irradiated sample.

This irradiation process leads to preferential C−H

bond scission. This implies that, in a well-oxygenated medium, alkyl peroxyl, ROO·, and hydroperoxyl, HOO·, radicals are produced. The latter are generally considered unable to participate in propagation: consequently, these radicals disappear essentially by termination reactions with ROO· radicals and production of ROOH. Even if this propagation reaction occurred, only the yield, and not the nature of the products obtained, would change.

The lack of carbonyl compounds among the products of the radiation-induced autoxidation of DMP at 25°C, (shown by IR spectroscopy) is therefore not surprising because the relative amount of β-scission of tertiary alcoxyl radicals compared to H-atom abstraction by these radicals decreases with the temperature. Thus β-scission can be neglected in the reaction scheme of the radiation-induced autoxidation of DMP at 25°C as well as the products of the initiation reactions, due to alkyl radical fragments resulting from the unfavored C–C bond radiolysis.

Reaction Scheme of the Radiation-Induced Oxidation of 2,4-Dimethylpentane at 25°C

Initiation

$$\text{H–R–H} \rightsquigarrow \text{H}^{\cdot}, \text{R}^{\cdot} \xrightarrow{O_2} \text{HOO}^{\cdot}, \text{H–R–OO}^{\cdot}\ \text{(I)} \quad (1)$$

Propagation by peroxyl radicals

$$\text{(I)} + \text{H–R–H} \xrightarrow{O_2} \text{HCO–R–H} + \text{(I)} \quad (2)$$

$$\textit{(intra)} \quad \text{(I)} \xrightarrow{O_2} \text{HOO–R–OO}^{\cdot}\ \text{(II)} \quad (2a)$$

$$\text{(II)} + \text{H–R–H} \xrightarrow{O_2} \text{HOO–R–OOH} + \text{(I)} \quad (3)$$

Non-terminating bimolecular reactions of peroxyl radicals

$$2\text{(I)} \longrightarrow 2\ \text{H–R–O}^{\cdot}\ \text{(III)} + O_2 \quad (4)$$

$$2\text{(V)} \longrightarrow 2\ \text{HO–R–O}^{\cdot}\ \text{(IV)} + O_2 \quad (5)$$

$$\text{(V)} + \text{(I)} \longrightarrow \text{(IV)} + \text{(III)} + O_2 \quad (5a)$$

Propagation by alcoxyl radicals

$$\text{(III)} + \text{H–R–H} \xrightarrow{O_2} \text{H–R–OH} + \text{(I)} \quad (6)$$

$$\textit{(intra)} \quad \text{(III)} \xrightarrow{O_2} \text{HO–R–OO}^{\cdot}\ \text{(V)} \quad (6a)$$

$$\text{(IV)} + \text{H–R–H} \xrightarrow{O_2} \text{HO–R–OH} + \text{(I)} \quad (7)$$

Secondary reactions [attack of HOO–R–H leading to HOO–R–OO· (II)]

$$\text{(I)} + \text{H–R–OOH} \xrightarrow{O_2} \text{H–R–OOH} + \text{(II)} \quad (8)$$

$$\text{(II)} + \text{H–R–OOH} \xrightarrow{O_2} \text{HOO–R–OOH} + \text{(II)} \quad (8a)$$

$$\text{(III)} + \text{H–R–OOH} \xrightarrow{O_2} \text{H–R–OH} + \text{(II)} \quad (8b)$$

$$\text{(IV)} + \text{H–R–OOH} \xrightarrow{O_2} \text{HO–R–OH} + \text{(II)} \quad (8c)$$

Termination

$$\text{(I)} + \text{HOO}^{\cdot} \longrightarrow \text{H–R–OOH} + O_2 \quad (9)$$

$$\text{(II)} + \text{HOO}^{\cdot} \longrightarrow \text{HOO–R–OOH} + O_2 \quad (10)$$

$$\left.\begin{array}{l}\text{(I)} + \text{(II)} \quad (11)\\ \text{(I)} + \text{(V)} \quad (12)\\ \text{(II)} + \text{(V)} \quad (13)\end{array}\right\} \longrightarrow \text{Dialkylperoxides} + O_2$$

3.1.2. EXPERIMENTAL RESULTS

The results are shown in Table 2 and Figures 1 and 2. The kinetics of production of the main products deduced

Table 2. Radiation-induced oxidation of DMP at 25°C, $I = 72$ rad min^{-1}.

Dose (Mrad) Time (hr)	0.207 48	0.415 96	0.622 144	0.972 225	1.188 275	1.404 325	1.732 401
[HO—R—H] (1)	ϵ	159	318	384	633	697	923
[HO—R—OH] (1)		480	596	1255	1484	1479	2061
[HO—R—H] (2)	327	698	1051	1536	1765	1860	2268
[HO—R—OH] (2)	235	500	748	1193	1637	2390	3187
Σ[products]	562	1198	1799	2729	3402	4250	5455
[HOO—R—H] (3)	327	539	733	1152	1132	1163	1345
[HOO—R—OOH] (3)		20	152		153	911	1126

[μmole l^{-1}]: (1) Before $\varnothing_3P$ reduction of hydroperoxides.
(2) After $\varnothing_3P$ reduction of hydroperoxides.
(3) Calculated.

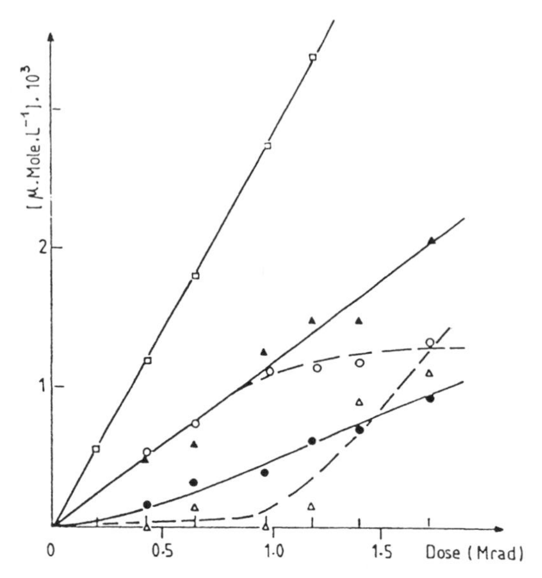

Figure 1. Radiation-induced oxidation of DMP at 25°C. □ Σ[products]; ○ [HOO−R−H]; △ [HOO−R−OOH]; ▲ [HO−R−OH]; ● [HO−R−H].

from the reaction scheme fit the experimental results quite well.

One can see in Figure 1 that intramolecular propagation reactions (by peroxyl radicals) are quite unfavoured [Reaction (2a)] and also that secondary reactions on HOO−R−H [Reactions (8)–(8c)] are negligible until the irradiation time reaches ~225 hr (~1 Mrad). This provides a relatively long initial period in which primary reactions can be easily studied.

3.2. Effects of sec.HAS Addition on the Oxidative Radiolysis of DMP at 25°C

3.2.1. *EFFECTS OF 2,2,6,6-TETRAMETHYLPIPERIDINE (2.95×10^{-2} mole l^{-1})*

The results are shown in Table 3 and Figures 3 and 4. Comparison with the above results shows:

(a) TMPH reduced the sums of the yields [HO−R−H] + [HOO−R−H] and [HO−R−OH] + [HOO−R−OOH]. But, in the initial period, the reduction in the accumulation rate of HO−R−H is very slight compared to that of HOO−R−H (Table 6).

The rate of accumulation of TMPO· is constant during the initial period and then decreases [the same kind of behavior will be observed during the thermal oxidation of LDPE at 95°C: the accumulation of >NO· increases first, reaches a maximum, and then decreases to a very low level (see § 4.1.2.)].

(b) The overall accumulation yields in the initial period of the radiation-induced oxidation [HO−R−H] + [HOO−R−H] + [HO−R−OH] + [HOO−R−OOH] + [>NO·] are directly proportional to the absorbed dose, as in the case of the oxidative radiolysis of pure DMP.

Furthermore, these overall yields are nearly equal. This observation and the fact that, in the initial period, the accumulation rate of >NO· is constant are incompatible with the production of >NO· resulting from a reaction between >NH and HOO−R−H [20-23]. Otherwise, this accumulation rate would correspond to that of a secondary product and would show an induction period.

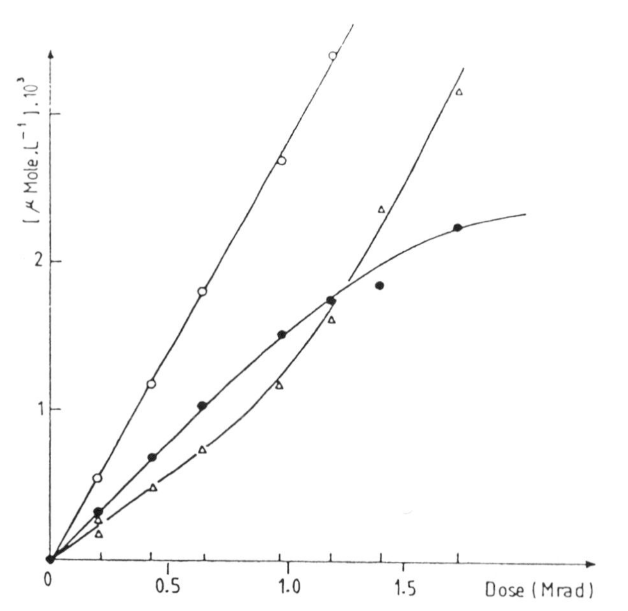

Figure 2. Radiation-induced oxidation of DMP at 25°C. ○ Σ[products]; ● [HO−R−H] + [HOO−R−H]; △ [HO−R−OH] + [HOO−R−OOH].

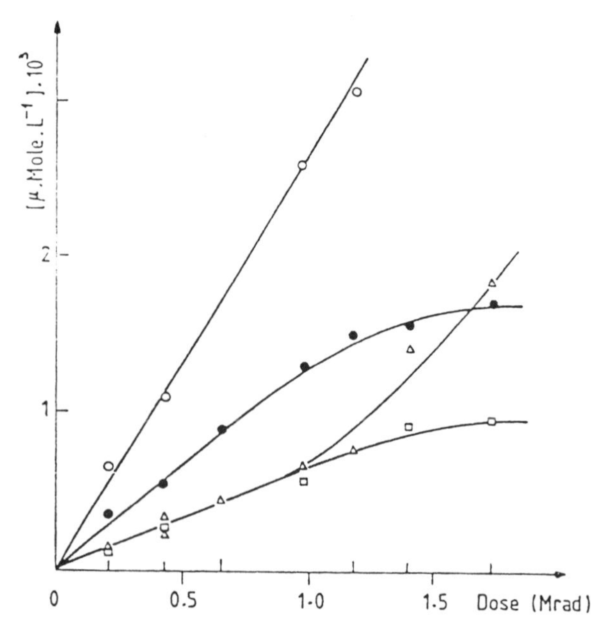

Figure 3. Radiation-induced oxidation of solutions DMP + TMPH (2.95×10^{-2} M) at 25°C. ○ Σ[products]; △ [HO−R−OH] + [HOO−R−OOH]; □ [TMPO·]; ● [HO−R−H] + [HOO−R−H].

Table 3. Radiation-induced oxidation of solutions DMP + TMPH (2.95×10^{-2} M) at 25°C, I = 72 rad min^{-1}.

Dose (Mrad) Time (hr)	0.207 48	0.415 96	0.622 144	0.972 225	1.188 275	1.404 325	1.732 401
[HO—R—H] (1)	128	192	335	362	446	598	649
[HO—R—OH] (1)	38	124		282	473	620	
[HO—R—H] (2)	345	543	900	1313	1516	1554	1704
[HO—R—OH] (2)	147	292	449	690	793	1438	1840
[TMPO·]	167	226		598	764	930	996
Σ[products]	659	1101		2601	3073	3922	4540
[HOO—R—H] (3)	217	351	565	951	1070	956	1055
[HOO—R—OOH] (3)	109	168		408	320	818	

[μmole l^{-1}]: (1) Before $\varnothing_3P$ reduction of hydroperoxides.
(2) After $\varnothing_3P$ reduction of hydroperoxides.
(3) Calculated.

This remark is also valid for the accumulation rate of HOO−R−H. If a reaction between HOO−R−H and >NH exists, a large induction period would be observed for hydroperoxides accumulation since the large excess of TMPH during the initial period would lead to a complete consumption of HOO−R−H. This is not the case (see Figure 4).

Therefore, it has been demonstrated that TMPO· is a primary product of the cooxidation of TMPH and DMP. Consequently, the production of TMPO· results from the competitive propagation reaction between TMPH and DMP in the consumption of free radical species, which are the precursors of the oxidation products of DMP. This reaction could be:

$$>NH \xrightarrow[RO\cdot]{ROO\cdot} >N\cdot \xrightarrow{O_2} >NOO\cdot \xrightarrow{>NOO\cdot} 2 >NO\cdot + O_2 \quad [24] \qquad (14)$$

$$>NOO\cdot + ROO\cdot \longrightarrow >NO\cdot + RO\cdot + O_2 \longrightarrow \text{propagation} \qquad (15)$$

This reaction chain explains why, in the initial period, the replacement of some ROO· by RO· does not cause significant decrease in HO−R−H yield, although the HOO−R−H yield decreases. The decrease in the yield of the non-terminating bimolecular Reaction (4) is compensated by Reaction (15).

3.2.2. EFFECTS OF TINUVIN 770 (3×10^{-2} mole l^{-1})

Results are shown in Table 4 and Figure 5. Comparison with the above results shows:

(a) The addition of Tinuvin 770, like that of TMPH, leads to lower sums of [HO−R−H] + [HOO−R−H] and [HO−R−OH] + [HOO−R−OOH]. But the slight decrease in HO−R−H yield which

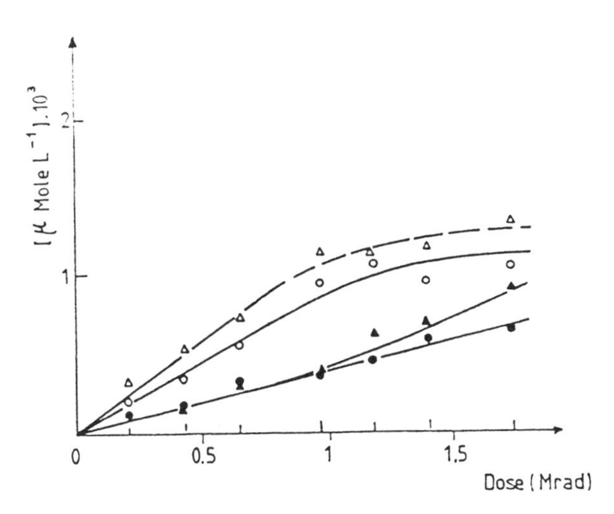

Figure 4. Radiation-induced oxidation of solutions DMP + TMPH (2.95×10^{-2} M) at 25°C.
a) DMP ▲ [HO−R−H] △ [HOO−R−H].
b) TMPH solution ● [HO−R−H] ○ [HOO−R−H].

Table 4. Radiation-induced oxidation of solution DMP + Tinuvin 770 (3.10 × 10^{-2} M) at 25°C, I = 72 rad min^{-1}.

Dose (Mrad) Time (hr)	0.207 48	0.415 96	0.622 144	0.972 225	1.188 275	1.404 325	1.732 401
[HO—R—H] (1)	59	165	224	360	426	450	597
[HO—R—OH] (1)	ϵ	ϵ	ϵ	95	404	935	1631
[HO—R—H] (2)	222	405	697	1089	1290	1536	1619
[HO—R—OH] (2)	322	734	1050	1720	2130	2709	
[·ON--NO·]		285	455	643	787		1488
Σ[products]	644	1424	2202	3452	4207		
[HOO—R—H] (3)	163	240	473	729	864	1086	1022
[HOO—R—OOH] (3)	322	734	1100	1625	1726	1774	

[μmole l^{-1}]: (1) Before $\varnothing_3P$ reduction of hydroperoxides. (2) After $\varnothing_3P$ reduction of hydroperoxides. (3) Calculated.

was observed can be explained by a less complete compensation of the two production reactions discussed above. The yield of HOO−R−OOH is much larger than for pure DMP or DMP + TMPH.

These two results may be interpreted with regards to the protective effect of reactive piperidyl groups which would result from the participation of diester bridges of Tinuvin 770 in the cooxidation of the system.

The accumulation rate of >NO· is not significantly modified within the limitations of the ESR measurements. This result is compatible with a compensation effect on [>NO·] which would be due to the piperidyl active groups' protection by diester bridges in the sec.HAS.

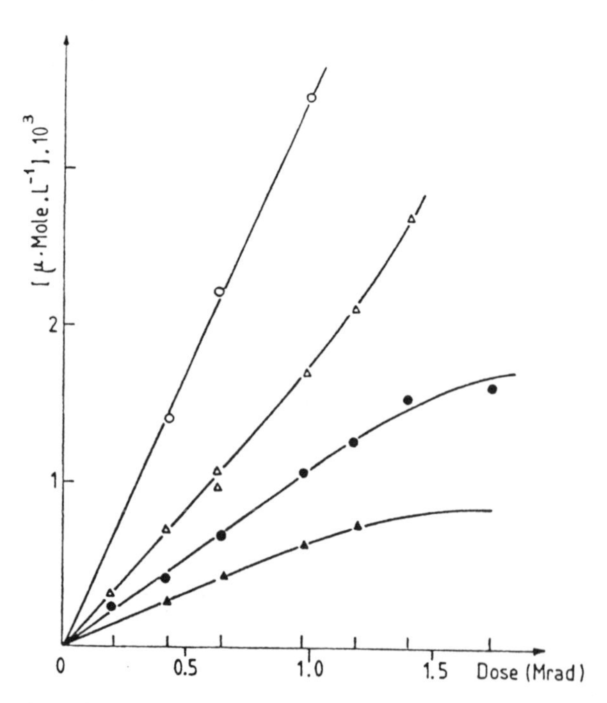

Figure 5. Radiation-induced oxidation of solutions DMP + Tinuvin 770 (3 × 10^{-2} M) 25°C. ○ Σ[products]; ● [HO−R−H] + [HOO−R−H]; ▲ [>NO·]; △ [HO−R−OH] + [HOO−R−OOH].

(b) HOO−R−OOH yield being higher, the overall yield [HO−R−H] + [HO−R−OH] + [HOO−R−H] + [HOO−R−OOH] + [>NO·], which remains proportional to the absorbed dose, is itself higher than corresponding overall yields observed during the oxidative radiolysis of DMP and of DMP + TMPH. Thus, it is even harder to imagine that this sec.HAS could be an AO (Figure 10).

3.3. Comparison with the Effects of a Phenolic AO: Ionol or BHT (2.96 × 10^{-2} mole l^{-1})

The results obtained with this well-known AO are shown in Table 5 and Figure 6. One can see that this AO, at equal active group concentrations, inhibits propagation reactions.

Figure 10 summarizes all the results by comparing the overall yields of the products of the oxidative radiolysis of DMP and of DMP containing each of the three additives discussed above. It clearly shows that the sec.HAS are not AO by themselves. Table 6 compares the initial

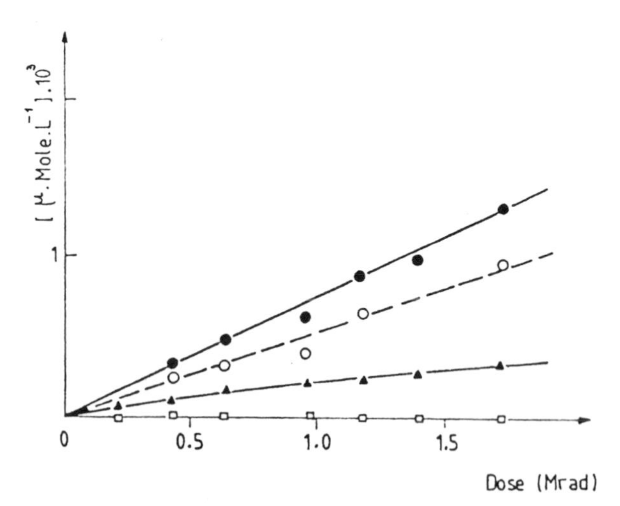

Figure 6. Radiation-induced oxidation of Ionol solutions DMP + Ionol (2.96 × 10^{-2} M). ● Σ[products]; ▲ [HO−R−H]; ○ [HOO−R−H]; □ [HO−R−OH] + [HOO−R−OOH].

Table 5. Radiation-induced oxidation of solution DMP + Ionol (2.96 × 10^{-2} M) at 25°C, I = 72 rad min^{-1}.

Dose (Mrad) / Time (hr)	0.207 / 48	0.415 / 96	0.622 / 144	0.972 / 225	1.188 / 275	1.404 / 325	1.732 / 401
[HO—R—H] (1)	47	89	159	219	228	287	340
[HO—R—OH] (1)	0	0	0	0	0	0	0
[HO—R—H] (2)	43	354	493	622	897	969	1336
[HO—R—OH] (2)	0	0	0	0	0	0	0
Σ[products]	43	354	493	622	897	969	1336
[HOO—R—H] (3)	0	265	334	403	669	682	996
[HOO—R—OOH] (3)	0	0	0	0	0	0	0

[μmole l^{-1}]: (1) Before $\varnothing_3P$ reduction of hydroperoxides.
(2) After $\varnothing_3P$ reduction of hydroperoxides.
(3) Calculated.

accumulation rates of all the products considered in Figure 10. These rates (in mole $l^{-1}h^{-1}$) were calculated by the least square method.

3.4. Conclusion

The major interest of these experiments on the radiation-induced oxidation of DMP at 25°C in the dark is that, using this method and GC analysis, it was possible to look at the primary reactions of sec.HAS in the initial period where secondary reactions are quite negligible.

The conclusion is that sec.HAS are co-reactive, the autooxidation of which competes with that of the substrate. *Consequently, it is not these additives themselves which act as AO, but the nitroxyl radicals >NO· which are their oxidation products. It is important to recognize this because, oxygenation of the medium being ensured during the experiments, the inhibition behavior of >NO· can no longer be explained by recombination with alkyl radicals or, presently, with H atoms.*

It will be shown (§ 5) that the nitroxyl radicals TMPO· and ·ON--NO· derived from the two studied sec.HAS are effectively strong AO able to inhibit the oxidative radiolysis of DMP.

The contradiction with the following results which will show the stabilization of LDPE conferred by sec.HAS against thermo-oxidative degradation, appears more disturbing than it is. At the beginning of the thermo-oxidation, an accumulation of >NO· will be observed which must be explained in the same way as that which occurred during the initial period of the radiation-induced oxidation of DMP at 25°C. But, this accumulation reaches a maximum and then decreases to a very low value which demonstrates that this species is consumed in the stabilization process. Taking into account the experimental conditions, these results will lead to the same conclusion—it is surely not the recombination of >NO· with alkyl radicals [25–30] nor H atoms which could explain their AO properties in an oxygen-saturated system.

4. KINETICS OF THE INHIBITION OF THE THERMO-OXIDATION OF LDPE FILMS BY sec. OR tert.HAS AT 95°C

4.1. Effects of Tinuvin 770 (Figure 7)

4.1.1. *IR SPECTROPHOTOMETRY*

The analysis of the products of the autoxidation of LDPE films, ketones and hydroperoxides, was already possible by IR spectrophotometry after 8 days of thermo-oxidation of the samples from the films ① and ①' (LDPE without additive) and after about 50 days for the samples from the industrial film ② (LDPE + 0.01 wt.% Irganox 1076). In all these samples, the accumulation of the products then increased very rapidly.

In the samples from films ④ (LDPE + 0.5 wt.% Tinuvin 770) and ⑤ (LDPE + 0.5 wt.% Tinuvin 770 + 0.01 wt.% Irganox 1076), the absorption bands corresponding to >N—H bond and to piperidyl groups of this sec.HAS (see § 2.5.) were no longer visible after 14 days of thermo-oxidation because a large amount of this additive evaporated at this temperature. The absorption peak

Table 6. Accumulation rates (mole l^{-1} hr^{-1}) of the products during the initial period of the autoxidation at 25°C.

	DMP	Ionol 2.96 × 10^{-2} M	TMPH 2.95 × 10^{-2} M	Tinuvin 770 3.00 × 10^{-2} M
HO—R—H	1.78 × 10^{-6}	0.84 × 10^{-6}	1.58 × 10^{-6}	1.47 × 10^{-6}
HOO—R—H	5.0 × 10^{-6}	2.3 × 10^{-6}	3.8 × 10^{-6}	3.2 × 10^{-6}
HO—R—OH	5.1 × 10^{-6}	0	1.3 × 10^{-6}	(0.4 × 10^{-6})
HOO—R—OOH	(0.6 × 10^{-6})	0	(1.1 × 10^{-6})	7.3 × 10^{-6}
>NO·			2.59 × 10^{-6}	2.89 × 10^{-6}

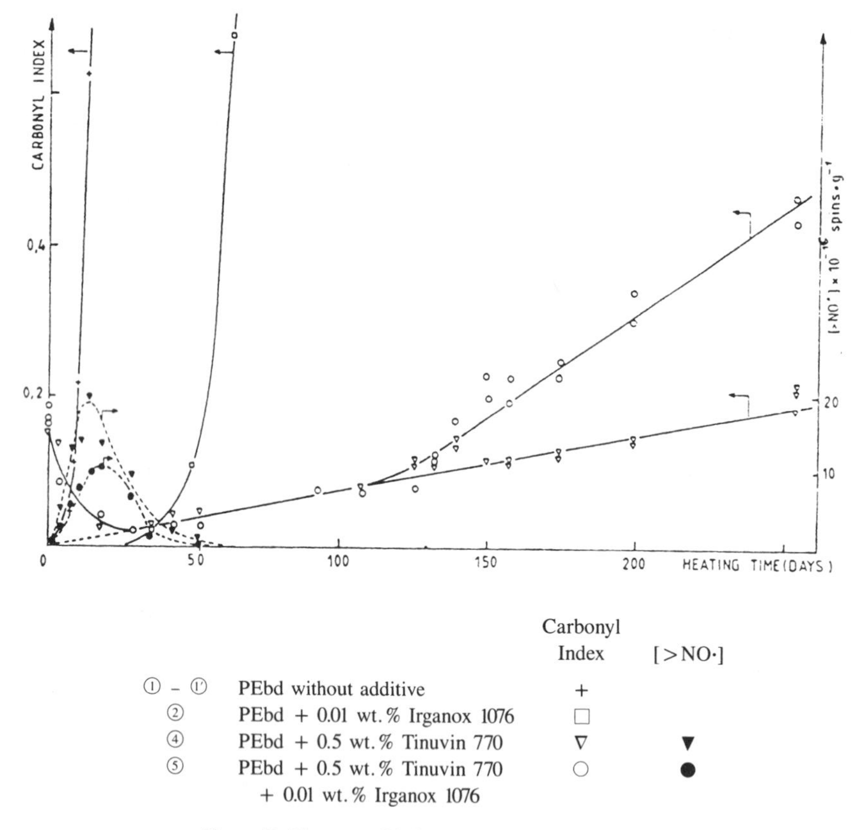

Figure 7. Thermo-oxidation of LDPE films at 95°C.

at 1718 cm^{-1}, initially due to the ester >C=O group, decreased rapidly during the first 20 days and reached a minimum. Then it increased slowly for the rest of the experiment, owing to the ketone >C=O groups produced by the autoxidation of the polymer.

In samples of film ④, the accumulation of functional groups absorbing in the region of carbonyl and carboxyl groups is quite slow compared to that of ① and ②. The absorption peak at 1718 cm^{-1} increases at a nearly constant rate from the beginning until the end of the experiments (254 days). In samples of film ⑤, the rate of accumulation of these functional groups increases after an initial period where it was nearly the same as that seen in samples of film ④.

One characteristic of samples ④ and ⑤ containing Tinuvin 770 is that no significant absorption in the region of O–H band due to hydroperoxides, alcohols... appeared, taking into account the limitations of the method.

4.1.2. ESR SPECTROMETRY

The kinetics of the [>NO·] accumulation in samples ④ and ⑤ are characterized by a rapid increase without induction period. Then, this concentration reaches a maximum and decreases quickly to very low values.

These results can be compared to those obtained during the photo-oxidation of polyolefins containing sec.HAS [2,15] and also those of Rozantsev et al. on the thermo-oxidation of polyisoprene containing 1,4-bis(2,2,6,6-tetramethylpiperidyl)butane reported by Shlyapintokh [2]:

$$H{-}N\langle\ \rangle{-}CH_2{-}CH_2{-}CH_2{-}CH_2{-}\langle\ \rangle N{-}H$$

The initial period of those kinetics requires no commentary because the accumulation of >NO· resulting from the cooxidation of sec.HAS agrees with their accepted unreactivity toward oxygenated radical species. But the interpretation of the latter part is completely different. *Therein lies the main interest for future research on sec.HAS stabilization because such kinetics demonstrate that a mechanism, unknown until now, consumes nitroxyl radicals during the autoxidation of an oxygen-saturated substrate.*

4.1.3 DISCUSSION: ANTAGONISM BETWEEN PROPERTIES OF sec.HAS AND PHENOLIC AO

The addition of 0.01 wt.% Irganox 1076, corresponding to only 1.9×10^{-4} mole kg^{-1} of active phenolic sites was enough to cause the increase of the accumulation rate of carbonyl and carboxyl groups in samples ⑤ as com-

Table 7. Radiation-induced oxidation of solution DMP + TMPO· (2.98×10^{-2} M) at 25°C, I = 72 rad min^{-1}.

Dose (Mrad) Time (hr)	0.207 48	0.415 96	0.622 144	0.972 225	1.188 275	1.404 325	1.732 401
[HO—R—H] (1)	0	53	94	151	185	241	241
[HO—R—OH] (1)	0	0	0	0	0	0	0
[HO—R—H] (2)	105	225	296	563	755	907	959
[HO—R—OH] (2)	0	0	0	0	0	0	ε
Σ[products]	105	225	296	563	755	907	959
[HOO—R—H] (3)	105	172	202	412	570	666	718
[HOO—R—OOH] (3)	0	0	0	0	0	0	0

[μmole l^{-1}]: (1) Before $\varnothing_3P$ reduction of hydroperoxides.
(2) After $\varnothing_3P$ reduction of hydroperoxides.
(3) Calculated.

pared to samples ④. This is evidence of the antagonism between the properties of these two stabilizers.

The lower maximum value of [>NO·] concentration observed during the thermo-oxidation of samples ⑤ confirms this antagonism. This behavior, already observed in photochemistry, results from the formation of thermally stable products by recombination of >NO· with cyclohexadienonyl radicals instead of the recombination of the latter with peroxyl radicals [26]. Another known reaction is the hydroxylamine production by phenolic hydrogen transfer to >NO·. But this last reaction does not change the overall production of >NO· because >NO· is quickly recovered by hydrogen transfer from >NOH to a propagative species.

This antagonism between Tinuvin 770 and Irganox 1076 in the stabilization of LDPE would be among the exceptions according to the conclusion of Allen et al. [31], recalled by Allen in a more general revue [26]. This conclusion states that in most cases synergy exists between the stabilization effects of sec.HAS and of phenolic AO in the thermo-oxidation of polyolefins while antagonism is more frequent during their photo-oxidation. Our opinion of this conclusion is nonetheless more cautious. On the basis of the proposed mechanisms of this antagonism, it may be obvious that secondary photochemical reactions can enhance the consequences of this antagonism by the initiation of new oxidative chain reactions of the substrate, but it is not obvious that these mechanisms could lead to synergy during thermo-oxidation.

4.2. Effects of Tinuvin 622

4.2.1. IR SPECTROPHOTOMETRY

No absorption in the region of O–H band which would have shown the production of hydroperoxides, alcohols... no noticeable change in the absorption band of ester carbonyl group (1736 cm^{1}) (see § 2.5.), no production of ketones (1718 cm^{-1}) was detectable during the entire thermo-oxidation of industrial samples ③ (LDPE + 0.5 wt.% Tinuvin 622 + 0.01 wt.% Irganox 1076) (254 days) and of laboratory samples ⑥ (LDPE + 0.5% Tinuvin 622) of which thermo-oxidation (107 days) was started later. *This leads to the evidence that tert.HAS are quite good AO.*

4.2.2. EPR SPECTROMETRY

No >NO· radical was detected during the thermo-oxidation of samples ⑥ and ③. Sometimes a weak signal seemed to appear but could not be resolved within experimental error.

The authors who observed these radicals all stress that their production from tert.HAS is much lower than from sec.HAS [2,32,33]. On the other hand, Felder [34] could not demonstrate their production during photo-oxidation

Table 8. Radiation-induced oxidation of solution DMP + ·ON--NO· (2.40×10^{-2} M) at 25°C, I = 72 rad min^{-1}.

Dose (Mrad) Time (hr)	0.207 48	0.415 96	0.622 144	0.972 225	1.188 275	1.404 325
[HO—R—H] (1)	ε	42	80	174	227	216
[HO—R—OH] (1)	0	0	0	0	0	0
[HO—R—H] (2)	101	222	395	599	816	
[HO—R—OH] (2)	46	42	17		35	129
Σ[products]	147	264	412	599	851	
[HOO—R—H] (3)	101	180	315	425	589	
[HOO—R—OOH] (3)	46	42	17	0	35	129

[μmole l^{-1}]: (1) Before $\varnothing_3P$ reduction of hydroperoxides.
(2) After $\varnothing_3P$ reduction of hydroperoxides.
(3) Calculated.

of isoctane containing 4-benzoyloxy-1-octyl-2,2,6,6-tetramethylpiperidine:

4.2.3. DISCUSSION: NO ANTAGONISM BETWEEN PROPERTIES OF tert.HAS AND PHENOLIC AO

The addition of 0.01 wt.% of Irganox 1076 did not modify the results. Therefore, there is apparently no antagonism between the AO properties of tert.HAS and of phenolic AO. Since tert.HAS are good AO, these results could also mean that their reaction products (unlike >NO· produced by reaction of sec.HAS) do not react with phenolic AO which would be protected when they are, as was the case in our experiments, in very weak concentrations compared to the tert.HAS. This would then be a *strong argument against >NO· production, at least by a primary reaction, during the stabilization by tert.HAS, the mechanism of which remains unsolved.*

5. INHIBITION OF THE RADIATION-INDUCED OXIDATIVE DEGRADATION OF DMP CONTAINING NITROXYL RADICALS TMPO· OR ·ON--NO·, AT 25°C

5.1. Experimental Results

The experiments, repeated under the conditions given in § 3—but after having replaced the sec.HAS by their nitroxyl derivatives, TMPO· and ·ON--NO· (at the same concentration ~3 × 10^{-2} mole l^{-1})—showed that *nitroxyl free radicals are AO that have the same effects as the phenolic AO previously used.*

The results are shown in Tables 7 and 8 and illustrated in Figures 8 and 9. The comparison of the results obtained with Ionol (Table 5, Figure 6) shows that the AO

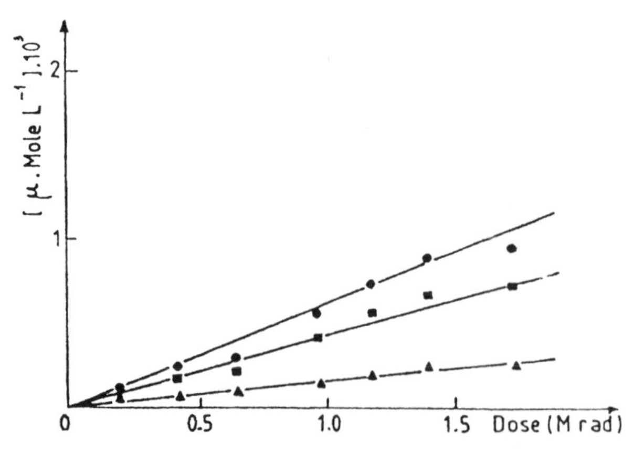

Figure 8. Radiation-induced oxidation of solutions DMP + TMPO· (2.98 × 10^{-2} M) at 25°C. ● Σ[products]; ■ [HOO–R–H]; ▲ [HO–R–H].

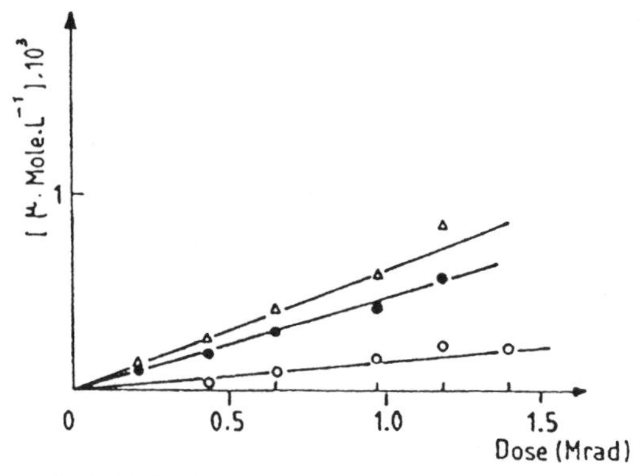

Figure 9. Radiation-induced oxidation of solutions DMP + ·ON--NO· (2.4 × 10^{-2} M) at 25°C. △ Σ[products]; ● [HOO–R–H]; ○ [HO–R–H].

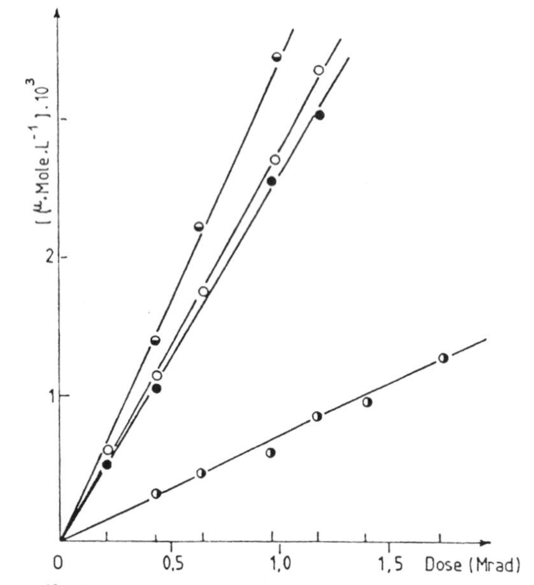

Figure 10. Overall yields Σ[products] in solutions DMP + TMPH or Tinuvin 770 or Ionol at 25°C. ○ DMP; ◓ Tinuvin 770 (3.0 × 10^{-2} M); ◑ Ionol (2.96 × 10^{-2} M); ● TMPH (2.95 × 10^{-2} M).

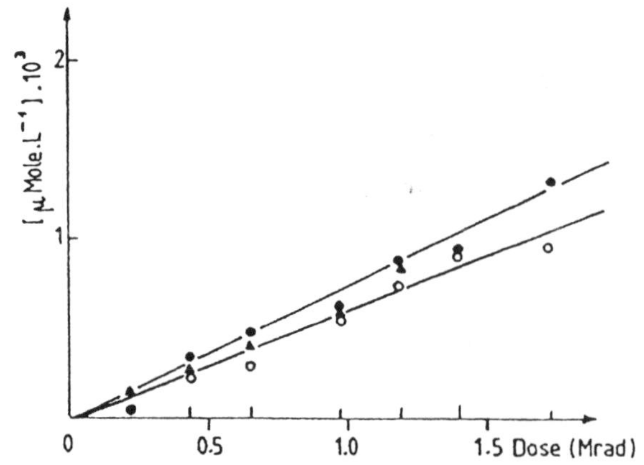

Figure 11. Overall yields Σ[products] in solution DMP + Ionol or TMPO· or ·ON--NO· at 25°C. ● Ionol (2.96 × 10^{-2} M); ▲ ·ON--NO· (2.4 × 10^{-2} M); ○ TMPO· (2.98 × 10^{-2} M).

Table 9. Accumulation rates (mole l^{-1} hr^{-1}) of the products during the initial period of the autoxidation at 25°C.

	DMP	Ionol 2.96×10^{-2} M	TMPO· 2.98×10^{-2} M	(·ON=NO·) 2.4×10^{-2} M
HO—R—H	1.78×10^{-6}	0.84×10^{-6}	0.66×10^{-6}	0.77×10^{-6}
HOO—R—H	5.0×10^{-6}	2.3×10^{-6}	1.72×10^{-6}	1.93×10^{-6}
HO—R—OH	5.1×10^{-6}	0	0	0
HOO—R—OOH	(0.6×10^{-6})	0	0	0

effects of TMPO· and of ·ON--NO· are even better at equal initial concentration of active sites, nitroxyl and phenolic (Figure 11). The decrease of the initial accumulation rates of HO−R−H and of HOO−R−H is a little larger, and as with Ionol the productions of HO−R−OH and HOO−R−OOH are no longer measurable (Table 9). The AO effects observed with TMPO· and with ·ON--NO· are practically equivalent, although with the latter, the concentration of >NO· active sites is twice that of TMPO·. This observation could mean a lowered AO effect for ·ON--NO·. This difference can be related to the participation (which has been shown) of diester bridges between piperidyl groups in the propagation of the cooxidation chain reaction of Tinuvin 770 and DMP.

5.2. Conclusion

These results confirm those obtained in the first part devoted to the cooxidation of sec.HAS and DMP (§ 3.4). They demonstrate that the protection against the oxidative degradation (which is induced by the addition of sec.HAS) is due to nitroxyl radicals >NO· resulting from their primary cooxidation reactions with the substrate.

Therefore, sec.HAS are not stabilizers by themselves but the generators of AO species which are their nitroxyl derivatives. The comparison of Figures 9 and 10 clearly illustrates this fact. *These results also confirm that nitroxyl radicals make excellent AO in oxygen-saturated systems, that is, under conditions where their recombination reaction with alkyl radicals (or H atoms) cannot compete with the peroxidation reaction of these radicals.* This is a problem because it is well accepted that the nitroxyl radical does not react with oxygenated radicals.

Further research, already started, must explore this problem in order to elucidate the AO mechanism of nitroxyl radicals.

REFERENCES

1. Sedlar, J., J. Marchal and J. Petruj. *Polym. Photochem.*, 2:175 (1982).
2. Shlyapintokh, V. Ya. and V. B. Ivanov. In *Developments in Polymer Stabilisation, Vol. 5*. G. Scott, ed. London: Applied Science, p. 41 (1982).
3. Klemchuk, P. P. *Ullmann's Encyclopedia of Industrial Chemistry, Vol. A3*. 5th ed. Weinheim:VCH Verlagsgesellschaft, p. 91 (1985).
4. Spinks, J. W. T. and R. J. Woods. *An Introduction to Radiation Chemistry*. 2nd ed. New York:Wiley-Interscience, p. 67 (1976).
5. Hilt, E., Thesis (ULP−Strasbourg) (1974). J.-C. Proquin, Thesis (ULP−Strasbourg) (1979).
6. Bagheri, R., K. B. Chakraborty and G. Scott. *Polym. Deg. and Stab.*, 4:1 (1982).
7. Chirinos Padron, D. J., F. A. Suarez and H. Berroteran. *Polym. Deg. and Stab.*, 14:295 (1986).
8. Petruj, J., S. Zehnacker, J. Sedlar and J. Marchal. *Polym. Deg. and Stab.*, 15:193 (1986).
9. Maier, R. D. and R. T. Hall. In *Organic Peroxides, Vol. 2*. D. Swern, ed. New York:Wiley-Interscience, p. 601 (1971).
10. Horner, L. and W. Jurgeleit. *Ann. Chem.*, 591:138 (1955).
11. Petruj, J., S. Zehnacker, J. Sedlar and J. Marchal. *The Analyst*, 111:671 (1986).
12. Carlsson, D. J. and D. M. Wiles. *Polym. Deg. and Stab.*, 6:1 (1984).
13. Allen, N. S., J. L. Kotecha, J.-L. Gardette and J. Lemaire. *Polym Deg. and Stab.*, 11:181 (1985).
14. Allen, N. S., J.-L. Gardette and J. Lemaire. *Polym. Deg. and Stab.*, 8:133 (1984).
15. Balint, G., A. Rockenbauer, T. Kelen, F. Tüdös and L. Jokay. *Polym. Photochem.*, 1:139 (1981).
16. Hodgeman, D. K. C. *J. Polym. Sci. Polym. Chem. Ed.*, 18:533 (1980).
17. Vink, P., R. T. Rotteveel and J. D. M. Wisse. *Polym. Deg. and Stab.*, 9:133 (1984).
18. Hodgeman, D. K. C. In *Developments in Polymer Degradation, Vol. 4*. N. Grassie, ed. London:Applied Science, p. 189 (1982).
19. Mill, T. and G. Montorsi. *Int. J. Chem. Kinet.*, 5:119 (1973).
20. Felder, B., R. Schumacher and F. Sitek. *Org. Coat. Plast. Chem.*, 42:561 (1980).
21. Felder, B., R. Schumacher and F. Sitek. In *Photodegradation and Photostabilisation of Coatings*. P. S. Pappas and F. H. Winslow, eds. ACS Symp. Series, Vol. 151, p. 65 (1981).
22. Chakraborty, K. B. and G. Scott. *Chem. Ind.*, 1:237 (1978).
23. Carlsson, D. J., A. Garton and D. M. Wiles. In *Developments in Polymer Stabilisation, Vol. 1*. G. Scott, ed. London:Applied Science, p. 219 (1979).
24. Faucitano, A., A. Buttafava, F. Martinotti and P. Bortolus. *J. Phys. Chem.*, 88:1187 (1984).

25. Scott, G., ed. *Developments in Polymer Stabilisation, Vol. 7*. London:Elsevier Applied Science, p. 65 (1984).

26. Allen, N. S. *Chem. Soc. Rev.*, 15:373 (1986).

27. Allen, N. S., ed. *Degradation and Stabilisation of Polyolefins*. London:Elsevier Applied Science (1984).

28. Gugumus, F. In *Developments in Polymer Stabilisation, Vol. 1*. G. Scott, ed. London:Applied Science, p. 261 (1979).

29. Wiles, D. M. and D. J. Carlsson. In *New Trends in the Photochemistry of Polymers*. N. S. Allen and J. F. Rabek, eds. London:Applied Science, p. 147 (1985).

30. Allen, N. S. In *New Trends in the Photochemistry of Polymers*. N. S. Allen and J. F. Rabek, eds. p. 209 (1985).

31. Allen, N. S., A. Hamidi, F. F. Loffelman, P. MacDonald, M. Rauhut and P. V. Susi. *Plant Rubb. Process Appl.*, 5: 259 (1985).

32. Kelen, T., F. Tüdös, G. Balint and A. Rockenbauer. *Am. Chem. Soc., Polym. Preprints*, 25:28 (1984).

33. Kurumada, T., H. Ohsawa, O. Oda, T. Fujita, T. Todo and T. Yoshioka. *J. Polym. Sci. Polym. Chem. Ed.*, 23:1477 (1985).

34. Felder, B. N. *Am. Chem. Soc., Polym. Preprints*, 25:26 (1984).

A. Factor[1]
W. V. Ligon[1]
R. J. May[1]
F. H. Greenberg[2]

Recent Advances in Polycarbonate Photodegradation

ABSTRACT

The mechanism of the photodegradation of bisphenol-A polycarbonate (BPA PC) was probed using state-of-the-art GC/GC/high resolution mass spectrometry, direct probe high resolution mass spectrometry, gas analysis studies on photolyzed ^{13}C labeled BPA PCs, and weathering of side chain free polycarbonates based on 3,3′-dihydroxydiphenyl ether. These studies show that the weathering of BPC PC involves multiple reaction pathways – photo-Fries reactions, side chain oxidation, ring oxidation and free radical ring attack. These results are consistent with a photo-aging mechanism in which side chain and ring photooxidation are initiated by a photo-Fries process resulting in chain cleavage and crosslinking reactions with ring oxidation being the primary source of colored photo-products.

KEY WORDS

Polycarbonate weathering, side chain free polycarbonates, isotopically labeled polycarbonates, photooxidation.

INTRODUCTION

The polycarbonate derived from bisphenol-A (BPA PC) is known for its toughness and transparency and is commonly used to replace glass in such applications as window glazing and light fixtures when glass breakage is a problem. However, when non-UV stabilized BPA PC is exposed outdoors for several years, its surface is observed to become yellow and show evidence of crosslinking, pitting and cracking. Most early investigations [1–8] on the photo-aging of BPA PC suggested that the key mechanism in this phenomenon is the photo-Fries reaction (Scheme 1). More recent work [9–16], however, indicates that under the influence of longer wavelengths of light – such as experienced during outdoor exposure – photooxidation (Scheme 2) is the predominant reaction pathway. In addition, work by Lemaire [13,14] demonstrated that photo-Fries products are themselves easily photooxidized making it difficult to find evidence of the photo-Fries pathway.

Although the evidence has been less rigorous, ring oxidation has been shown to play a role in the photo-aging of BPA PC. For example, ESCA studies of resin photooxidized under both sunlight ($\lambda > 300$ nm) [11] and Hg arc light ($\lambda > 280$ nm) [10,12] show a decrease of the $\pi \rightarrow \pi^*$ shake-up ESCA satellite indicating the loss of aromatic groups. Also, oxygen uptake experiments [9] indicate the absorption of up to 12 moles of O_2 per monomer unit and the evolution of 4.6 to 7.7 moles of $CO + CO_2$ per monomer unit during the photooxidation of thin polymer films using different light sources. These high values cannot be explained by side chain oxidation and carbonate group hydrolysis alone and indicate that ring oxidation is also occurring.

While the above work has done much to explain the origins of the photo-aging of BPA PC, a number of important questions remain unanswered, such as the identity of the yellow photoproducts, the nature of the crosslinking reactions, and the relative importance of side chain vs. ring oxidation in photoyellowing. In an effort to clarify these points, three experimental approaches were undertaken.

The first approach entailed the synthesis and weathering of polycarbonates based on bisphenols similar to BPA but free of alkyl groups – namely 3,3′-dihydroxydiphenyl ether (m-BPO) (I) and 4,4′-dihydroxydiphenyl ether (p-BPO) (II) [17(a)]. The hitherto unknown polycarbonate from m-BPO was specifically chosen as the key bisphenol unit in this study both because it does not contain

HO OH

m-BPO

I

[1]General Electric Corporate Research & Development Center, Schenectady, NY 12301.

[2]Department of Chemistry, Buffalo State College, Buffalo, NY 14222.

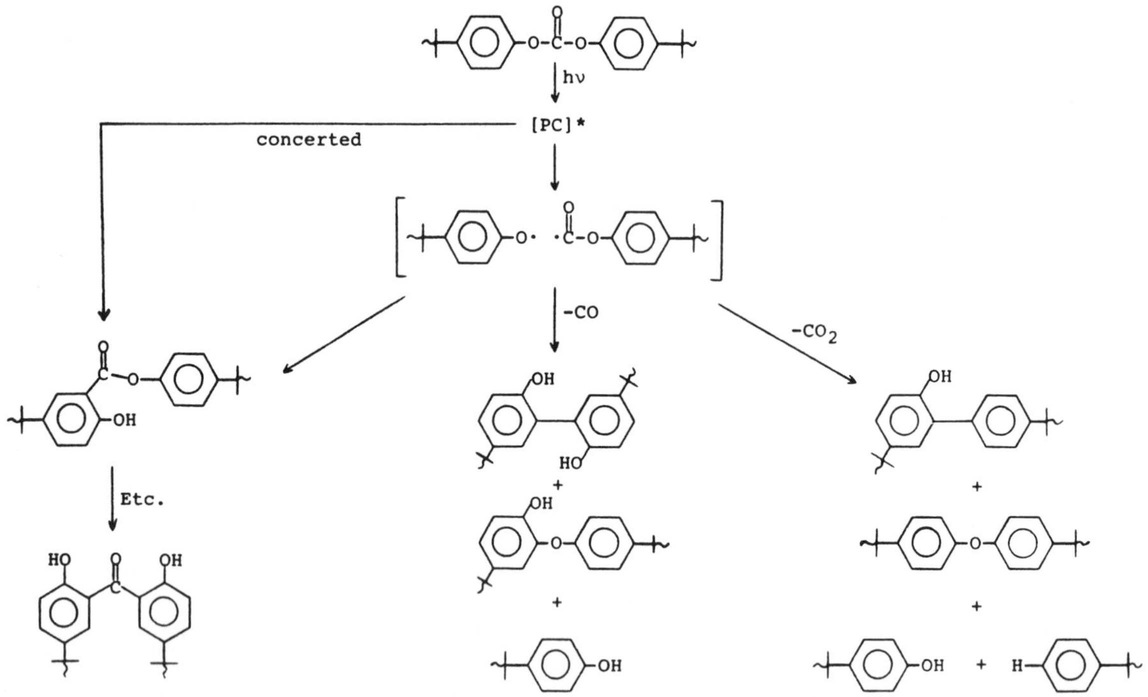
hν
[PC]*
concerted
-CO
$-CO_2$
Etc.

Scheme 1. Photo-Fries pathway.

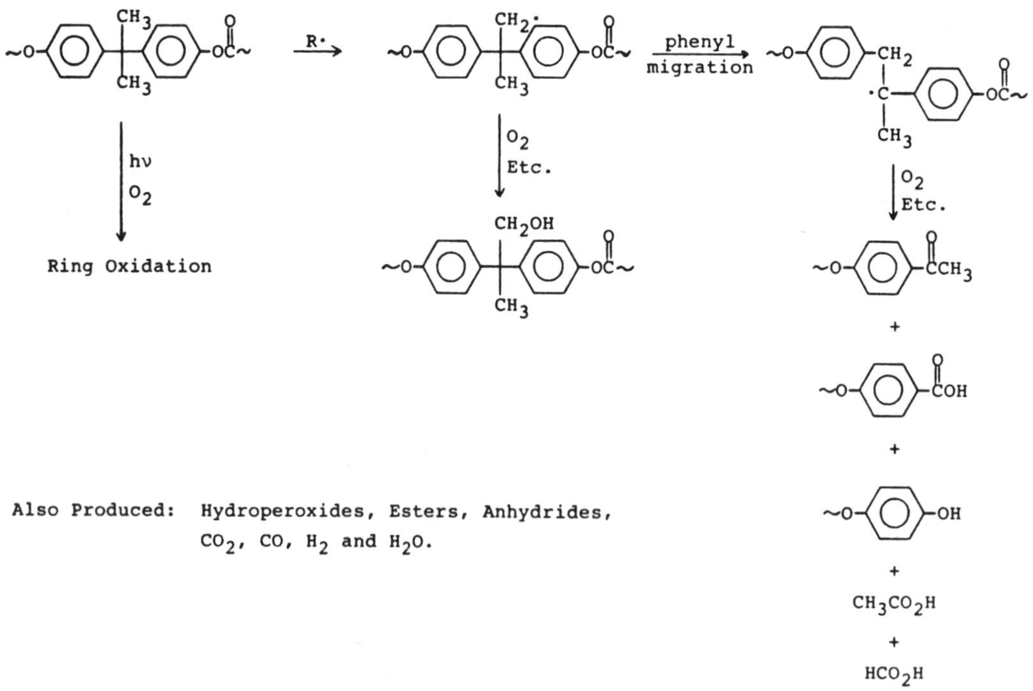
R·
phenyl
migration
hν
O_2
Ring Oxidation
O_2
Etc.
O_2
Etc.
CH_3CO_2H
HCO_2H
Also Produced: Hydroperoxides, Esters, Anhydrides,
CO_2, CO, H_2 and H_2O.

Scheme 2. Photooxidative pathway.

HO–C6H4–O–C6H4–OH (p-BPO, II)   HO–C6H4–C(Me)(Me)–C6H4–OH (BPA, III)

p-BPO BPA

II III

methyl groups as in BPA (III) and because its meta substitution pattern precludes the formation of the colored p-quinone during weathering and diminishes the chances that m-BPO PC will be highly crystalline, a problem encountered in the synthesis of p-BPO PC [18].

In the second approach, an in-depth analytical study was made of a photo-aged, non-UV stabilized commercial sample of BPA PC using GC/GC/high resolution mass spectrometry (GC/GC/Hi-Res MS) and solids probe high resolution mass spectrometry [17(b)].

The third approach involved the photooxidation in a sealed system of BPA PCs isotopically labeled with ^{13}C at either the carbonate group or the ortho positions of the BPA rings. By assaying the amounts of $^{13}CO_2$ and ^{13}CO evolved during this treatment, one can in theory assay the relative importance both of the photo-Fries vs. photooxidation reactions and of side chain oxidation vs. ring oxidation reactions.

EXPERIMENTAL

Materials

SYNTHESIS OF 3,3'-DIHYDROXYDIPHENYL ETHER (I)

The route to m-BPO (I), shown in Equation (1), is based on the Ullman aryl ether synthesis. The experimental details are given elsewhere [17(a)].

MeO–C6H4–OH + Br–C6H4–OMe —1) Ullmann Rx, 2) HI→ HO–C6H4–O–C6H4–OH (1)

SYNTHESIS OF POLYCARBONATES

Polycarbonates were prepared by phosgenation using standard aqueous caustic conditions [17(a)], shown in Equation (2).

HO–C6H4–X–C6H4–OH + $COCl_2$ —aq. NaOH, Et_3N→ [–O–C6H4–X–C6H4–OC(=O)–] (2)

BISPHENOL-A POLYCARBONATE

The bisphenol-A polycarbonate sample used in the GC/GC/Hi-Res MS study was a non-UV stabilized Lexan®* 101–112 grade polycarbonate in the form of a 4" × 1/8" molded disc which had been weathered 4 years in Florida. The resin had an $[\eta]$ of 0.57 dl/g ($CHCl_3$), was endcapped with phenol, and contained 0.1 wt.% of a phosphite/epoxy heat stabilizer system. Samples of the photo-aged surface were isolated by manually scraping the yellowed surface with a curved scalpel to obtain thin (~10 μm) yellow shavings of polymer. A sample of resin cut from the interior of the disc was used as an unreacted control.

Lithium aluminum hydride (LAH), 95+% purity and *lithium aluminum deuteride* (LAD), 95 atom % D, were purchased from the Aldrich Chemical Company.

5-α,α(Dimethyl-p-hydroxybenzyl)-salicylic acid (o-carboxy bisphenol-A) was purchased from the Aldrich Chemical Company (Cat. No. 14,9098).

2-(4-Hydroxyphenyl)-2-(3,4-dihydroxyphenyl)propane was prepared by the transalkylation method of Mark [19], Equation (3).

HO–C6H4–C(CH3)2–C6H4–OH + catechol (C6H4(OH)2) ⇌(H^+) HO–C6H4–C(CH3)2–C6H3(OH)–OH + HO–C6H5 (3)

A mixture of 0.506 g bisphenol-A (2.2 mmol), 0.233 g catechol (2.0 mmol) and three drops of methane sulfonic acid were heated with stirring in the melt at 125°C for 30 min. GC analysis showed the presence of a number of reaction products including bisphenol-A and a large amount of phenol. The desired product was separated and analyzed by GC/GC/Hi-Res MS.

2,2-Bis(p-hydroxyphenyl)propanol was prepared in analytical quantities by the above LAH reduction procedure from 2,2-bis(p-hydroxyphenyl)propionic acid synthesized by the method of Parris, et al. [20].

^{13}C Labeled Bisphenol-A Polycarbonates

BPA PC with >90% ^{13}C enrichment at the carbonyl position was kindly supplied to us by Dr. P. M. Henricks [21]. BPA PC with >90% ^{13}C enrichment at one of the two ortho carbons in each ring was generously provided by Drs. P. T. Inglefield and A. Jones [22]. The positions and percent of ^{13}C enrichments were confirmed by ^{13}C NMR spectroscopy.

*Lexan® is a registered trademark of the General Electric Co.

Procedures

PHOTOAGING OF SIDE CHAIN FREE POLYCARBONATES

Polymers were weathered as CH_2Cl_2 cast films both in a Q-Panel company QUV Weathering Tester using a cycle "8 hrs. at 70°C light/4 hrs. at 50°C dark condensation" and by direct outdoor exposure in Schenectady, NY (45° mounting facing south). The color of the films was monitored in yellowness index (*YI*) units measured with a Gardner Model XL-20 colorimeter using the transmission operation mode.

REDUCTIVE HYDROLYSIS PROCEDURE

Theoretically, three moles of LAH are required to reductively cleave four moles of an aromatic carbonate as shown in Equation (4).

$$3\ LiAlH_4 + 4\ ArOC(=O)OAr \xrightarrow{THF} \xrightarrow[\text{work up}]{H_2O} 4\ CH_3OH + 3\ Al(OH)_3 + 3\ LiOAr + 5\ HOAr \quad (4)$$

In practice, about a 4-fold excess of LAH (or LAD) was utilized to compensate for the presence of adventitious H_2O. Typically about 30 mg of polycarbonate (0.12 mmoles) and about 20 mg of LAH (0.53 mmoles) in 2 ml of dry, purified THF were magnetically stirred overnight in a closed vial to form a thick white paste which was worked up by first adding 20 μl of distilled H_2O (1.11 mmoles) to the stirred slurry followed by the addition of several small pieces of dry ice to neutralize the reaction products. To remove metal salts, the resulting slurry was filtered through a small plug of silica gel in a glass pipette and eluted with dry methanol. The trimethylsilyl derivatives of the resulting products were prepared by treating the evaporated solutions with N,O-bis(trimethylsilyl)trifluoroacetamide.

ANALYTICAL PROCEDURES

GC/GC/Hi-Res MS experiments were performed using a Vg Analytical ZAB mass spectrometer system which has been described in detail previously [23]. In this work, the first stage of GC separation was accomplished using a 10 ft. by 2 mm i.d. GC column packed with 3% OV1 on 100/120 mesh Gas Chrom Q (Applied Science Laboratories). Samples of about 2 μl volume (50% solids in bis-trimethylsilyltrifluoroacetamide) were injected at a column temperature of 70°C, and the column was then temperature programmed from 70–290°C at 10°C per minute. Ten percent of the effluent from the primary column was split to a flame ionization detector (FID) to obtain a primary chromatogram. Using this primary chromatogram, both the solvent peak and the vast majority of the BPA peak were vented from the instrument while the rest of the column effluent was diverted to a cold trap and then re-injected onto a second GC column for further separation.

The second stage of GC separation was accomplished by using a J&W Scientific Model DB-1 fused silica capillary column. The column had a 0.25 μm film thickness and 0.25 mm i.d. The contents of the cold trap were injected at 70°C, and the oven was temperature programmed from 70–290°C at 8°C per minute. Both stages of the GC separation utilized helium for the carrier gas. The fused silica capillary was connected directly to the ionization volume of a high resolution mass spectrometer, and no molecular separator was employed.

The ion source was operated in electron impact mode at 70 eV electron energy and a trap current of 270 μA. During the course of the secondary chromatogram, mass spectra were acquired continuously at a resolution of 10,000 ($m/\Delta m$, 10% valley definition). Spectra were obtained over the mass range 200–500 amu using 2.6 second upscans with an interscan delay of 0.4 seconds. These conditions provided 11 ADC samples per peak at the data system and a mass measurement accuracy consistently better than 2.0 millimass units. Low resolution mass spectral data ($m/\Delta m$ 1000) were also acquired over the mass range 20–600 amu which provided full fragmentation patterns which were used to help make structural assignments. Data was acquired using a Finnigan-MAT model 2400C mass spectrometry data system running under "SUPERINCOS" software. Perfluorokerosene was admitted continuously with the sample to allow high resolution mass measurement.

For the solids probe experiments, a one microliter aliquot of each of the silylated samples was dispersed on prebaked pyrex glass powder and put into a small crucible. The crucible was placed in the solids probe of the Vg ZAB mass spectrometer, and the probe was slowly heated from 50° to 250°C. During the evaporation, the instrument was scanned repetitively from 300 to 1000 amu at a rate of 10 seconds/decade using a mass resolution ($m/\Delta m$) of 10,000. Perfluorokerosene was admitted together with the sample to allow precise mass measurements. Ionization was obtained by electron ionization and the ion source temperature was 230°C. At first, a large amount of silylated BPA was observed to evaporate followed by a smaller quantity of higher molecular weight materials. Spectra obtained after most of the BPA had evaporated were summed to obtain an average spectrum, and the precise mass assignments were made using this averaged data.

GAS ANALYSIS OF PHOTOOXIDIZED ISOTOPICALLY LABELED BPA PC

These experiments were carried out as previously described [9] in sealed Pyrex glass flasks containing a 90/10 mixture of O_2 and Ar. Samples of ~ 10 μ films (~20 mg) were exposed for 2 months on each side to the light from a 275 watt GE RS sunlamp while the sample temperature was maintained at 26° ±2°C using a water bath. At the end of the exposure, the gas compositions in the reaction flasks were determined using high resolution mass spectrometry.

RESULTS AND DISCUSSION

Photo-Aging of Side Chain Free Polycarbonates

POLYMER CHARACTERIZATION

The polymerization of m-BPO and m-BPO/p-BPO systems proceeded without any problems using the same conditions routinely employed for preparing BPA PC. The structure of the resulting polymer was confirmed by ^{1}H NMR, ^{13}C NMR, and IR spectra. The physical properties measured for these polymers are reported in Table 1.

The UV spectrum has similar λ_{max} values to that of BPA PC with peaks at 264 nm (shoulder), 269 nm, and 276 nm compared to the spectrum of BPA PC [24], which has peaks at 265 nm and 272 nm. However, as indicated in Table 1, it was found that the m-BPO PC is twice as absorbing as the BPA PC. TGA analysis of m-BPO PC indicates a high degree of stability, with weight loss occurring at 405°C (N_2) and 415°C (air). The (η) of the m-BPO PC was 0.20 dl/g suggesting that the polymer might be of low molecular weight. GPC analysis shows the material consists of a broad molecular weight range peaking at M_w = 15,700 g/mol (polystyrene standard) with a number of low molecular weight components comprising ~12% of the peak area. Reverse phase LC analysis (THF/H_2O) confirmed the presence of a series of low molecular weight linear and cyclic m-BPO PC oligomers. The reason for the formation of high concentration of cyclics (and linear oligomers) is presently not well understood but it is likely due to the bent U shape of the m-BPO (I) and the flexible nature of its aromatic ether linkage which favor the formation of cyclics. This also explains the DSC analyses which indicate that the T_g of the m-BPO PC was 71.1°C, ~80°C lower than the T_g of BPA PC.

COMPARATIVE WEATHERING BEHAVIOR

The UV behavior of m-BPO PC, 1/1 m-BPO/p-BPO PC, and the BPA PC was assessed on thin films (~2.5 mil) using both a QUV accelerated weathering apparatus and outdoor exposure. The results, shown in Table 1, show that after both exposures the change of yellowness index (ΔYI) was greatest for the m-BPO/p-BPO copolymer and least for the BPA PC.

That a side chain free aromatic polycarbonate such as m-BPO PC photoyellows at all establishes that ring oxidation is an important source of photoyellowing. The fact that these side chain free aromatic polycarbonates photoyellow even faster than the side chain containing BPA PC suggests that ring oxidation may be even more important than side chain oxidation in the formation of yellow photoproducts. The increased light absorption and the higher concentration of the more photoreactive uncapped linear oligomers in m-BPO containing PCs certainly contribute to the faster photoyellowing found in these systems. To better gauge the importance of these factors, future model studies are planned to compare the rates of photooxidation and photoyellowing of the bisphenyl carbonates of m-BPO and BPA.

High Resolution Mass Spectrometric Analysis of Weathered BPA PC

Reductive cleavage of the photo-aged polycarbonate was performed using both LAH and LAD so that one could simultaneously cleave the carbonate linkages in the polymer and isotopically label any carbonyl containing groups in the resin [refer to Equations (5) and (6)].

$$\mathrm{R\overset{O}{\overset{\|}{C}}OR'} \xrightarrow[\text{2) } H_2O]{\text{1) LAD}} \mathrm{RCD_2OH} + \mathrm{HOR'} \quad (5)$$

Table 1. Properties of m-BPO based polycarbonates.

System	25° η $CHCl_3$ (dl/g)	T_g (°C)	TGA Temp. at 10% Wt. Loss Air	N$_2$	UV λ_{max} (ϵ)	QUV Weathering[a] ΔYI @ 71 Hrs.	Outdoor Weathering[b] ΔYI @ 11 Weeks
m-BPO PC	0.20	71	415°	405°	264 nm (s) (1870), 269 nm (2058), 276 nm (1831)	14.7	1.9
1/1 m/p-BPO PC	0.35	84	440°	450°	264 nm (s) (1980), 270 nm (2129), 275 nm (2043)	22.9	4.8
BPA PC	0.55	148	482°	502°	265 nm (870), 272 nm (1000)	4.0	0.6

[a]Exposure cycle: 8 hrs. UV 70°C/4 hrs. dark condensation 50°C.
[b]Exposed 9/10/84 to 11/27/84 in Schenectady, NY.

$$\mathrm{R\overset{O}{\overset{\|}{C}}R'} \xrightarrow[\text{2) } H_2O]{\text{1) LAD}} \mathrm{R\overset{OH}{\overset{|}{C}}DR'} \tag{6}$$

Thus, any compound found to contain two Ds is derived from a carboxylic acid or ester, whereas those containing one D are derived from a ketone or aldehyde.

The capillary GC of the product mixture is shown in Figure 1, and a summary of the high resolution mass spectral analysis of the peaks in Figure 1 is listed in Table 2.

Those compounds also present in the control sample are so indicated in Table 2. As can be seen, nearly three dozen products of bisphenol-A polycarbonate were identified, most of which had never been previously found. Structural determinations were made using standard mass spectrometric procedures. In those cases, where standards or library spectra were available, structural assignments were made with relatively high confidence. When a molecular ion and abundant fragments were both present, it was often possible to make very reasonable assignments even in the absence of standards. This was greatly facilitated by the universal application of high resolution measurements which provided elemental compositions. However, as is typical for aromatic compounds, many components provided mass spectra consisting only of a molecular ion and few fragments. In such cases, speculative structures have been proposed based solely on elemental composition and on a knowledge of the chemistry involved.

Along with each structural designation, an attempt was made to assign the mechanistic pathway by which each compound was formed, e.g., ring oxidation (RO), ring attack (RA), side chain oxidation (SCO), and photo-

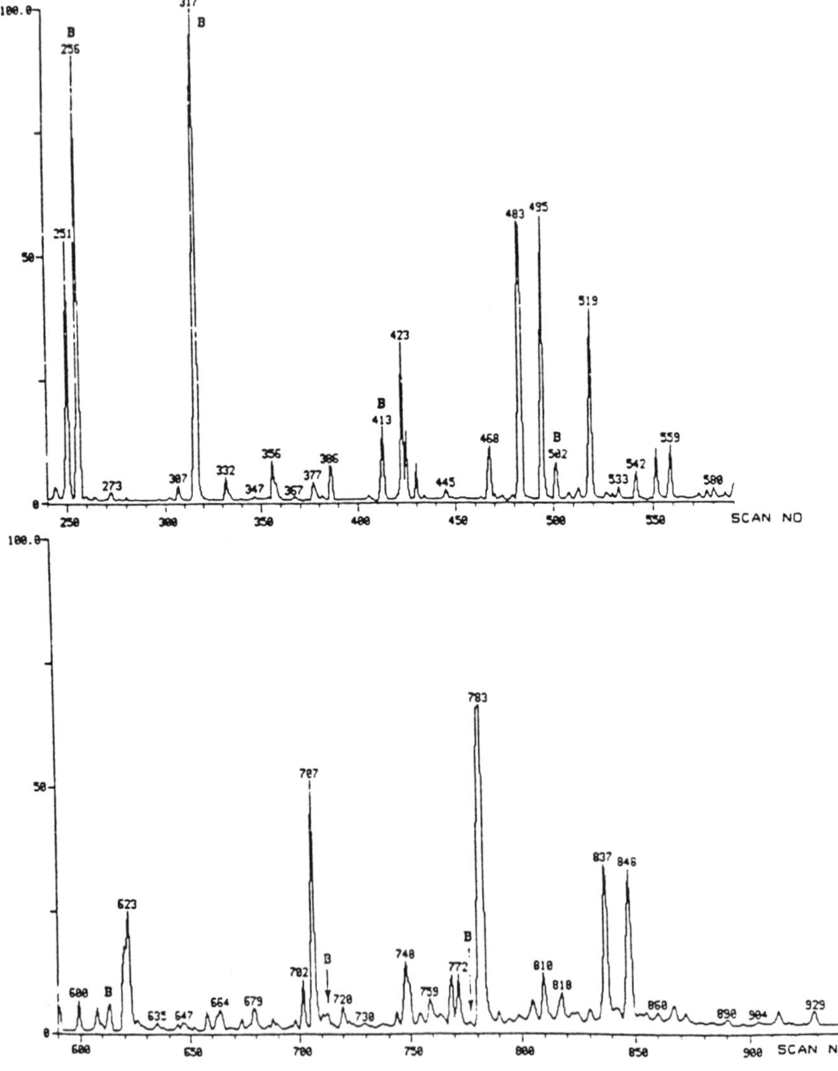

Figure 1. Capillary gas chromatogram of the product mixture. (Peaks labeled "B" are due to primary column bleeding of stationary phase.)

Table 2. High resolution mass spectral results from weathered bisphenol-A polycarbonate.

Scan[a] No.	Most Abundant MS Spectral Peaks[b]	Parent Mass ($M^{\dagger}$)	Parent Peak Hi-Res Formula[c]	Relative Yield[d]	Structural Assignment[e]	Likely Formation Pathway[f] RO	RA	SCO	PF	Probable Source/Comments
250[h]	(low res)	166	$C_9H_{14}OSi$	High	TMSO–⌬					End Cap
307	(low res)	180	$C_{10}H_{16}OSi$	Low	TMSO–⌬–Me (para)		X			Matches Authentic Sample
356	(low res)	194	$C_{11}H_{18}OSi$	Low	TMSO–⌬ (C_2H_5)			X		
377	(low res)	192	$C_{11}H_{16}OSi$	Low	TMSO–⌬–CH=CH$_2$			X		
386	208	208.1288	$C_{12}H_{20}OSi$	Low	TMSO–⌬–CH(CH$_3$)$_2$		X			
423a[h]	206	206.1122	$C_{12}H_{18}OSi$	High	TMSO–⌬–C(CH$_3$)=CH$_2$		X			
423b	209,224	224.1230	$C_{12}H_{19}DO_2Si$	Med.	TMSO–⌬–C(D)(CH$_3$)OCH$_3$			X		~O–⌬–C(=O)CH$_3$
445	239,254	254.1154	$C_{12}H_{22}O_2Si_2$	Low	TMSO–⌬–OTMS			X		
468	223,206,207,238	238.1404	$C_{13}H_{22}O_2Si$	Med.	TMSO–⌬–C(CH$_3$)$_2$OCH$_3$		X			~O–⌬–C(CH$_3$)=CH$_2$
483	267,282,251	282.1977	$C_{14}H_{25}DO_2Si_2$	High	TMSO–⌬–C(OTMS)(CH$_3$)D			X		~O–⌬–C(=O)CH$_3$
495	268,267,253,223, 237	268.1322	$C_{13}H_{22}D_2O_2Si_2$	High	TMSO–⌬–CD$_2$OTMS			X		~O–⌬–CO$_2$H
519	281,207,206,296, 265	296.1597	$C_{15}H_{28}O_2Si_2$	High	TMSO–⌬–C(CH$_3$)$_2$OTMS			X		
552	267,223,282	282.1102	$C_{13}H_{22}O_3Si_2$	Med.	TMSO–⌬–C(=O)OTMS			X		Incomplete Reduction

(continued)

Table 2. (continued).

Scan[a] No.	Most Abundant MS Spectral Peaks[b]	Parent Mass (M^{+})	Parent Peak Hi-Res Formula[c]	Relative Yield[d]	Structural Assignment[e]	Likely Formation Pathway[f] RO	RA	SCO	PF	Probable Source/Comments
559	207,223,295,310	310.1770	$C_{16}H_{28}D_2O_2Si_2$	Med.	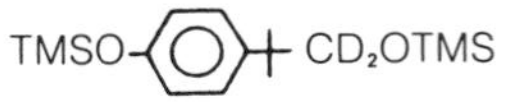 TMSO … CD$_2$OTMS	X				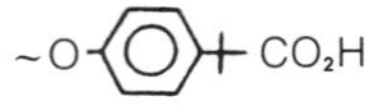 ~O … CO$_2$H
608a	207,257,272,213	272.1610	$C_{17}H_{23}DOSi$	Low	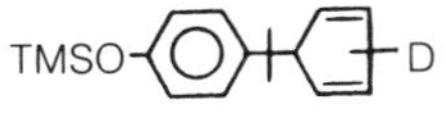 TMSO … D	X				Speculative Assignment
608b	207,324,223,253, 309	324.1953	$C_{17}H_{32}O_2Si_2$	Low	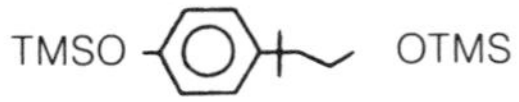 TMSO … OTMS	X				Speculative Assignment
623	281,294,206,265	294.1461	$C_{15}H_{26}O_2Si_2$	Med.	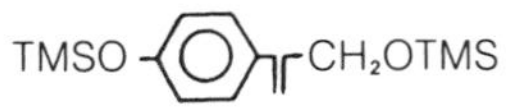 TMSO … CH$_2$OTMS			X		Prob. a mixture. Speculative Assignment.
658	308,293,219,267	308.1631	$C_{16}H_{26}D_2O_2Si_2$	Low	 TMSO … CH$_2$CD$_2$OTMS	X				Speculative Assignment
664	269,284	284.1573	$C_{18}H_{24}OSi$	Low	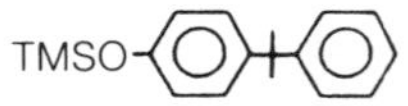 TMSO …				X	
679	207,205,309,233	324.1909	($C_{21}H_{27}DOSi$)[g]	Low	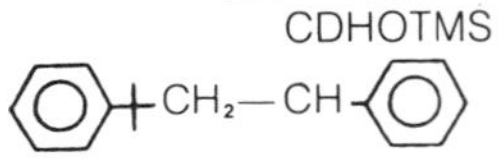 … CH$_2$—CH(CDHOTMS) …			X		Likely from IPP dimer impurity
707	231,246,257,319, 272,347,362,335, 219,207	362.2087	$C_{20}H_{33}DO_2Si_2$	High	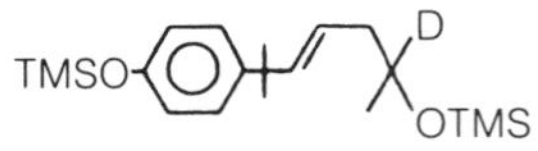 TMSO … D, OTMS	X				Speculative Assignment
713	207,231,362,244	362.2079	$C_{20}H_{34}O_2Si_2$	Low	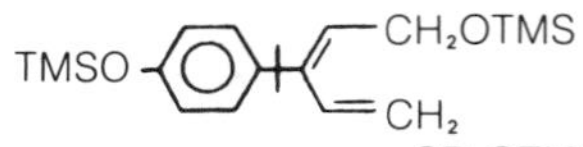 TMSO … CH$_2$OTMS, =CH$_2$	X				Speculative Assignment
744	207,348,267,245	348.1957	$C_{19}H_{30}D_2O_2Si_2$	Low	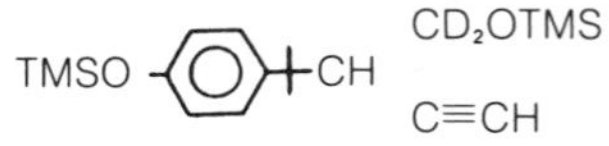 TMSO … CH(CD$_2$OTMS)(C≡CH)	X				Speculative Assignment
748[h]	207,358,372,334	372.1935	$C_{21}H_{32}O_2Si_2$	Med.	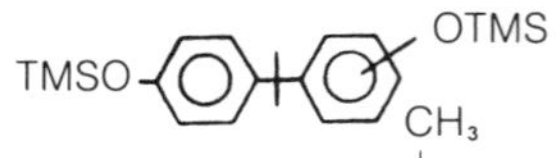 TMSO … OTMS					o,p-BPA also found in blank
759a	357,372,344,267	372.1941	$C_{21}H_{32}O_2Si_2$	Low	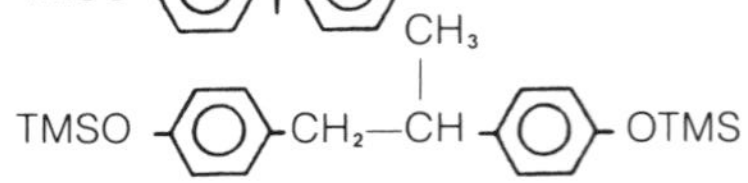 TMSO … CH$_2$—CH(CH$_3$) … OTMS			X		Speculative Assignment
759b	343,358	358.1741	$C_{20}H_{30}O_2Si_2$	Low	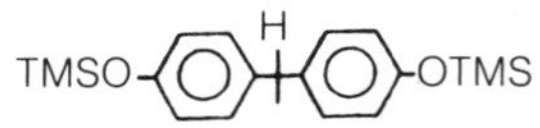 TMSO … (H) … OTMS			X		

(continued)

Scan[a] No.	Most Abundant MS Spectral Peaks[b]	Parent Mass (M^{+})	Parent Peak Hi-Res Formula[c]	Relative Yield[d]	Structural Assignment[e]	Likely Formation Pathway[f] RO	RA	SCO	PF	Probable Source/Comments
783[h]	357,372,207,341, 387	372.1931	$C_{21}H_{32}O_2Si_2$	High	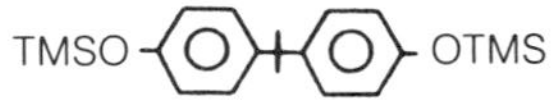 TMSO–(C₆H₄)–C(CH₃)₂–(C₆H₄)–OTMS					Residual BPA
810[h]	397,207,348,412, 231	412.2258	$C_{24}H_{36}O_2Si_2$	Low	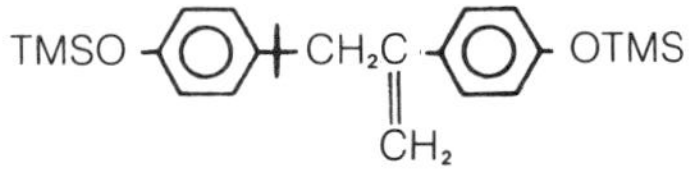 TMSO–(C₆H₄)–C(CH₃)₂–CH₂C(=CH₂)–(C₆H₄)–OTMS					IPP Dimer also found in blank
818	371,267,207,386, 297	386.2090	$C_{22}H_{32}D_2O_2Si_2$	Low	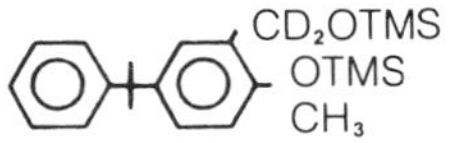 (C₆H₅)–C(CH₃)₂–(C₆H₃)(CD₂OTMS)–OTMS				X	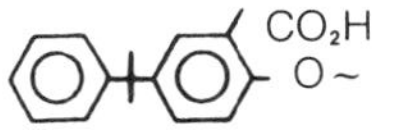 (C₆H₅)–C(CH₃)₂–(C₆H₃)(CO₂H)–O~
830	445,460,207,370, 223	460.2278	$C_{24}H_{40}O_3Si_3$	Low	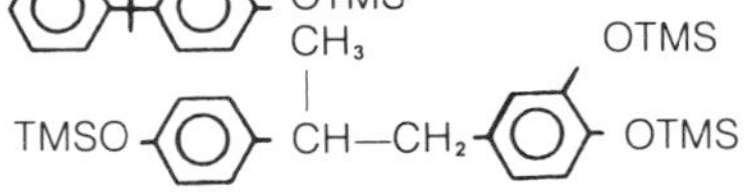 TMSO–(C₆H₄)–CH(CH₃)–CH₂–(C₆H₃)(OTMS)–OTMS	X		X		Speculative Assignment
837a	445,460,279,429, 295	460.2277	$C_{24}H_{40}O_3Si_3$	High	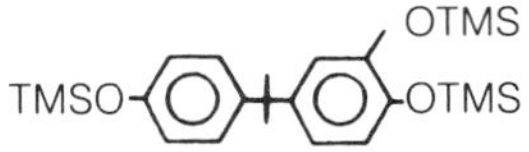 TMSO–(C₆H₄)–C(CH₃)₂–(C₆H₃)(OTMS)–OTMS	X				Matches Authentic Sample
837b	279,207,223,473, 488	488.2588	$C_{26}H_{43}DO_3Si_3$	Low	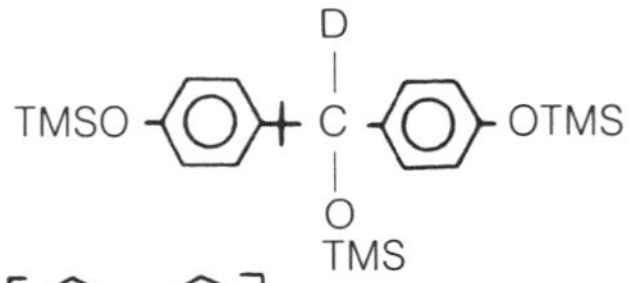 TMSO–(C₆H₄)–C(CH₃)₂–C(D)(OTMS)–(C₆H₄)–OTMS			X		Via IPP Dimer
837c	386,371	386.1733	$C_{21}H_{30}O_3Si_3$	Low	 [fused ring structure]–(OTMS)₃	X				Via TO–(dimethylfluorene)–OT ? 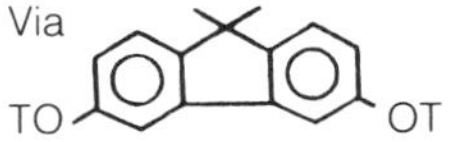
848	357,281,341,371, 445,460	460.2285	$C_{24}H_{40}O_3Si_3$	High	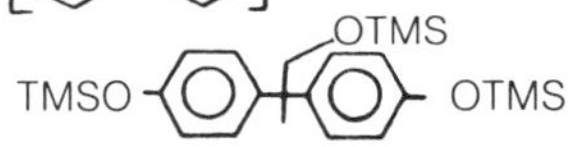 TMSO–(C₆H₄)–C(CH₃)₂–(C₆H₃)(OTMS)–OTMS			X		Matches Authentic Sample
867	459,207,281,267, 474	474.2439	$C_{25}H_{40}D_2O_3Si_3$	Low	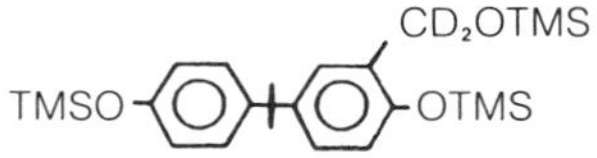 TMSO–(C₆H₄)–C(CH₃)₂–(C₆H₃)(CD₂OTMS)–OTMS				X	Matches Authentic Sample
872	370	370.1769	$C_{21}H_{30}O_2Si_2$	Low	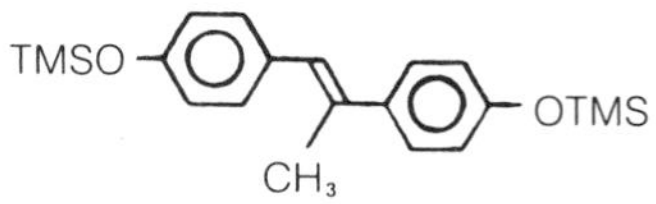 TMSO–(C₆H₄)–CH=C(CH₃)–(C₆H₄)–OTMS			X		Speculative Assignment

(continued)

Table 2. (continued).

Scan[a] No.	Most Abundant MS Spectral Peaks[b]	Parent Mass (M^{+})	Parent Peak Hi-Res Formula[c]	Relative Yield[d]	Structural Assignment[e]	Likely Formation Pathway[f] RO	RA	SCO	PF	Probable Source/Comments
904	473,488	488.2594	$C_{26}H_{42}D_2O_3Si_3$	Low	TMSO-C6H4-CH(CH3)CH2CH(CD2OTMS)-C6H4-OTMS			X		Speculative Assignment From IPP Dimer
913	357,223,398,474, 459	474.2446	$C_{25}H_{42}O_3Si_3$	Low	TMSO-C6H4-C(R)-C6H4-OTMS; R = C_2H_4OTMS			X		357 Strongest Peak due to: TMSO-C6H4-C⊕-C6H4-OTMS (913 and 929a)
929a	357,223,343,488	488.2609	$C_{26}H_{44}O_3Si_3$	Low	TMSO-C6H4-C(R)-C6H4-OTMS; R = C_3H_6OTMS			X		
929b	487,502	502.2764	$C_{27}H_{46}O_3Si_3$	Low						

[a]GC peaks due to more than a single component are designated with subscripts, a, b, c, etc.
[b]In the cases of spectra with >200 amμ peaks, only >200 amμ peaks reported.
[c]If peak from LAD treatment contains D, then D containing empirical formula given.
[d]Key: High (>50% of peak 495), Medium (20–49% of peak 495), Low (<20% of peak 495).
[e]Structures assigned by correspondence of MS with that in computer library and/or logical deduction from MS fragmentation pattern and most likely pathways of formation. TMS corresponds to a trimethyl silyl group.
[f]Key: RO—ring oxidation; RA—ring attack; SCO—side chain oxidation; PF—photo-Fries.
[g]Determined from parent peak minus 15 Hi-Res measurement.
[h]Compound also present in the control obtained from the interior of the photo-aged sample.

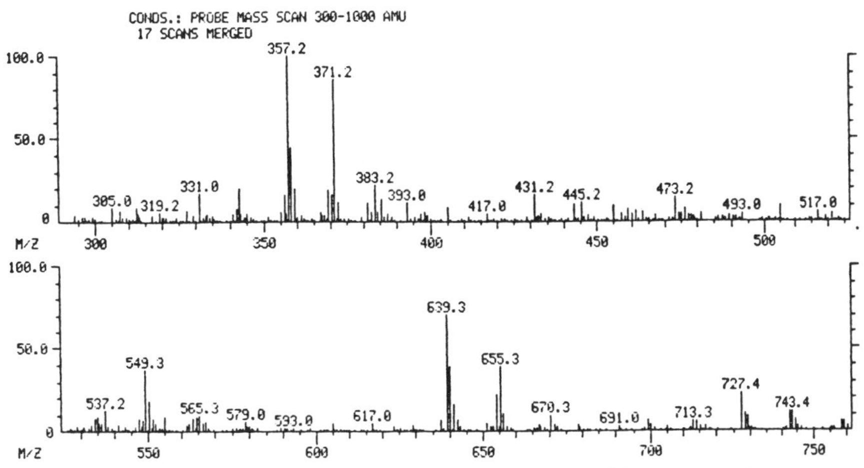

Figure 2. High resolution mass spectrum obtained using solids probe technique.

Fries (PF) (see Schemes 1 and 2). The ring attack pathway has not previously been considered important but appears necessary to explain the formation of several of the alkylated phenols identified, e.g., Equation (7). A similar mechanism has been postulated to explain the formation of 2,4,6-trimethyl phenol during the pyrolysis of poly(2,6-dimethyl-1,4-phenylene) ether [25].

(7)

Overall, the results shown not only confirm the previous work showing the importance of side chain oxidation in outdoor aged bisphenol-A polycarbonate [9,10,13]; but, for the first time they present hard evidence for the occurrence of the photo-Fries and ring oxidation during natural weathering. For example, the formation of the compounds analyzed in scans nos. 664, 818, and 867 are best rationalized as being formed via a photo-Fries reaction, whereas compounds analyzed in scans nos. 559, 608a, 608b, 658, 707, 713, 744, 830, 837a, 837c and 872 are best explained by a ring oxidation pathway. Admittedly, many of the structural assignments made of these products are only speculative (and are so indicated), especially those compounds we propose as arising from ring oxidation. Nonetheless, a number of assignments have been confirmed by comparison with authentic samples.

An important limitation of the GC/GC/Hi-Res MS technique is that it will not detect compounds with insufficient volatility to pass through the gas chromatographs. In order to assay for the presence of such materials, silylated samples of reductively cleaved polymer were analyzed by directly introducing them into the ionization chamber of the high resolution mass spectrometer using a solids probe which was then slowly heated from 50° to 250°C. Initially, a large amount of silylated BPA was detected followed by the appearance of a mixture of higher molecular weight compounds. The spectra thus obtained were summed to obtain an average spectrum (Figure 2), and as shown in Table 3, precise mass assignments were made using this averaged data. Although this technique simultaneously gives the mass spectra of all the components in the product mixture, one can easily recognize the parent and p-15 peaks of at least five compounds. While it is at present not possible to unambiguously identify these compounds, reasonable assignments could be made (cf. Table 3) based on either the photo-Fries processes shown in Scheme 3 and/or the occurrence of oxidative coupling of phenolic end groups.

The occurrence of such photo-Fries processes has previously been suggested by Pryde [15] and Webb [16] and is in accord with both the above GC/GC/Hi-Res MS results and the ^{13}C carbonyl labeled studies (see below) and in part accounts for the branching and crosslinking which occurs during polycarbonate weathering.

Another limitation of the above mass spectral work is

Table 3. High resolution mass spectral data from solids probe analysis.

Parent Mass ($M^{\ddagger}$)	Parent Peak Hi-Res Formula	Structural Assignment	Probable Source
1. 564.2869	$C_{32}H_{48}O_3Si_3$	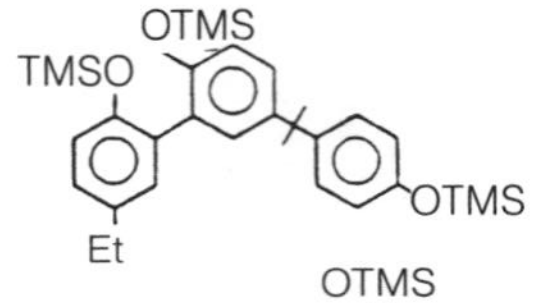 OTMS, TMSO, OTMS, Et	Photo-Fries plus Side Chain Oxidation
2. 654.3342	$C_{39}H_{54}O_3Si_3$	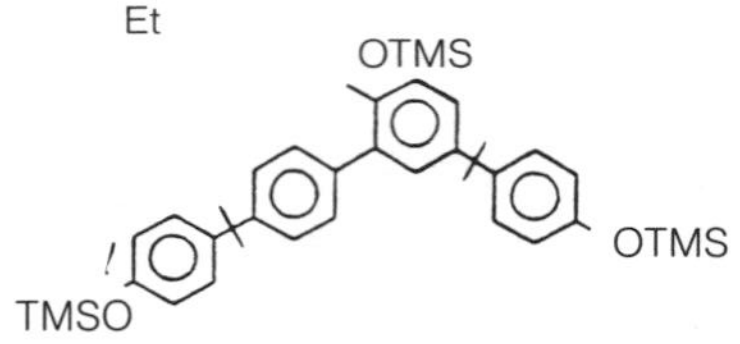 OTMS, OTMS, TMSO	Photo-Fries
3. 670.3320	$C_{39}H_{54}O_4Si_3$	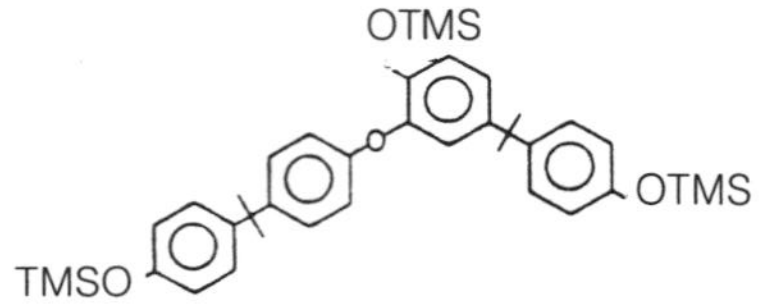 OTMS, OTMS, TMSO	Photo-Fries
4. 742.3711	$C_{42}H_{62}O_4Si_4$	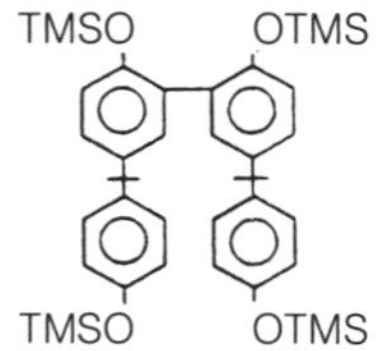 TMSO, OTMS, TMSO, OTMS	Photo-Fries
5. 758.3706	$C_{42}H_{62}O_5Si_4$	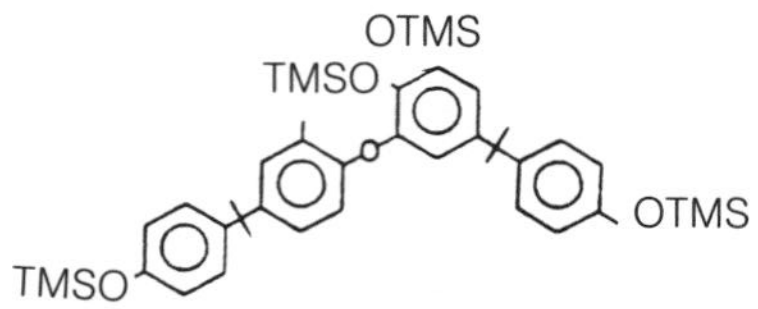 OTMS, TMSO, OTMS, TMSO	Photo-Fries and Ring Oxidation

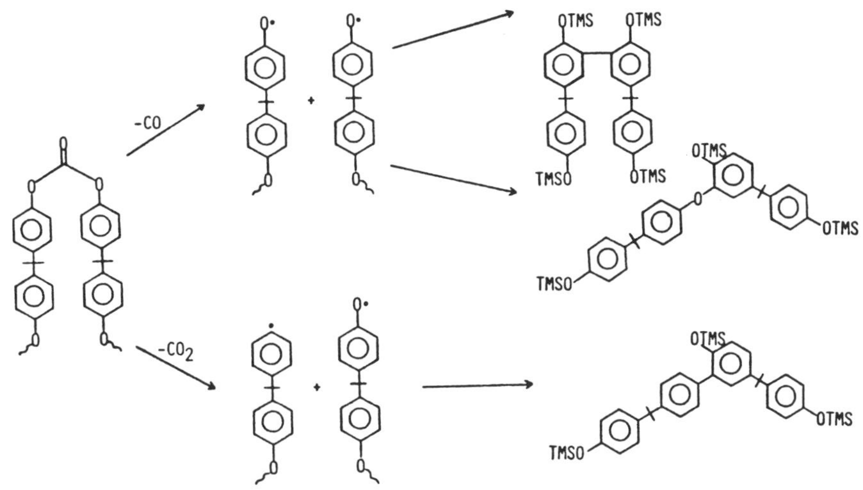

Scheme 3. Formation of high molecular weight products via the photo-Fries reaction.

the qualitative nature of the results. In an effort to obtain more quantitative evidence on the relative importance of the photo-Fries vs. photooxidation reactions and of ring oxidation vs. side chain oxidation reactions, sealed bulb experiments were carried out using two ^{13}C labeled BPA PC polymers—one with >90% ^{13}C in the carbonyl position and the other with >95% ^{13}C in one of the ortho positions in each phenyl ring. Thin (~10 μ) films of each polymer were sealed in 50 ml round Pyrex flasks under ~0.9 atm of a 90 O_2/10 Ar atmosphere and exposed both to sunlight and to the light of a 275 watt RS sunlamp (λ > 280 nm) at ~26°C.

The key measurement in these experiments is the percent of ^{13}C label found in the CO and CO_2 formed during weathering. In the case of carbonyl labeled BPA PC, ^{13}CO can only be from the photo-Fries reaction while ^{12}CO can only be derived from photooxidation. The presence of $^{13}CO_2$ is less diagnostic for it can arise from both the photo-Fries process and hydrolysis of the carbonate group. In contrast, in the case of the ring labeled BPA PC, both ^{13}CO and $^{13}CO_2$ are diagnostic of ring oxidation. While the outdoor exposure experiments are still in progress, the results using accelerated weathering conditions are reported in Table 4.

The CO values from carbonyl labeled BPA PC experiment indicate that under the exposure conditions used at least 5% of the photochemistry involved photo-Fries process. Since the ring labeled BPA PC contains ~12.5% ^{13}C, the high value of $^{13}CO_2$ found demonstrates that at least under the exposure conditions used, ring oxidation is as important a process as side chain oxidation. The lower value found for ^{13}CO in this case indicates that the processes leading to CO formation are more than twice as likely to occur during side chain oxidation plus photo-Fries reactions than during ring oxidation, i.e. (12.5 − 3.7/12.5 = 0.704 vs. 3.7/12.5 = 0.296.

As suggested by the above m-BPO PC weathering results, the occurrence of ring oxidation best explains the yellow discoloration that always occurs during the photooxidation of bisphenol-A polycarbonate. While specific darkly colored compounds have not as yet been identified, the empirical formulas obtained for many of these "ring oxidation" products indicate the presence of a high degree of reactive unsaturated groups which could be easily oxidized to colored products. In addition, it is well-known that multi-hydroxylated benzene systems such as the catechol derivative of scan 837a are quite prone to further oxidation to darkly colored products.

Table 4. ^{13}C labeled BPA PC experiments (4 months' exposure RS sunlamp/Pyrex under O_2 at 25°C).

	Yield CO_2/CO	Isotopic Yield		Isotopic Yield	
		^{12}CO	^{13}CO	$^{12}CO_2$	$^{13}CO_2$
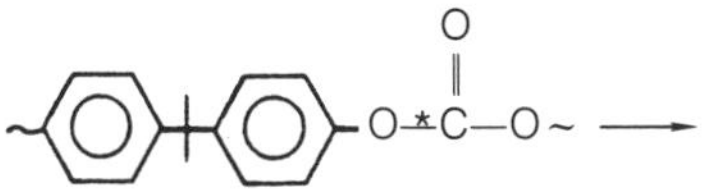 ~Ph–C(CH3)2–Ph–O–*C(=O)–O~ →	—	95.4%	4.6%	86.3%	13.7%
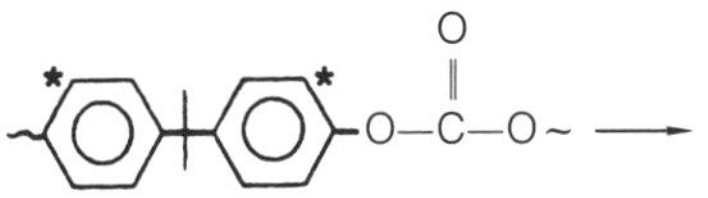 ~*Ph–C(CH3)2–Ph*–O–C(=O)–O~ →	3.8/1	96.3%	3.7%	88.7%	11.3%

Table 5. The effects of the four processes involved in polycarbonate weathering.

	Chain Scission	Crosslinking	Color Formation
Photo-Fries	+ + +	+ +	+
Side Chain Oxidation	+ + +	+	o
Ring Oxidation	+	+	+ + +
Ring Attack	+	o	o

Key: + + + Large Effect
+ + Medium Effect
+ Small Effect
o No Effect

The finding of compounds which are best explained by a photo-Fries pathway is of particular mechanistic interest. Previously, we postulated [9] that the photo-Fries reaction, while not a major pathway in the outdoor aging of bisphenol-A polycarbonate, is a key contributor during the early stages of the reaction before the concentration of more photo-reactive products becomes significant. The fact that evidence for the occurrence of the photo-Fries process was not previously found in outdoor aged samples [9,11,15] can be explained by the work of Lemaire [13,14] who demonstrated that photo-Fries products, if formed, were themselves easily photo-degraded. Thus, the finding of trace amounts of these products is in accord with the role of radical initiator of this reaction in the early stages of bisphenol-A polycarbonate weathering.

CONCLUSIONS

The above studies show that photochemistry resulting from the exposure of BPA PC to ultraviolet light and O_2 involves multiple reaction pathways, i.e., photo-Fries reactions, side chain oxidation, ring oxidation and free radical ring attack. As indicated in Table 5, the combined effects of these processes can account for the chain scission, crosslinking and photoyellowing observed in weathered polycarbonates.

ACKNOWLEDGEMENTS

The authors would like to thank Mr. R. O. Carhart of the General Electric Plastics Business Group for kindly supplying the sample of Florida aged bisphenol-A polycarbonate which analysis is the basis of this work. In addition, thanks are due to Drs. S. H. Schroeter and J. E. Pickett for illuminating discussions and advice and to Messrs. S. B. Dorn and H. Grade for mass spectrometric gas analysis. Lastly, the authors want to express their appreciation to Mrs. C. A. Herderich for her patience and expertise in the preparation of this paper.

REFERENCES

1. Bellus, D., P. Hrdlovic and Z. Manasek. *Poly. Letters*, 4:1 (1966).
2. a) Davis, G. A. and J. H. Golden. *J. Chem. Soc.*, (B):426 (1968); b) G. A. Davis, J. H. Golden, J. A. McRae and M. C. R. Symons. *Chem. Comm.*, 398 (1967); and c) J. A. McRae and M. C. R. Symons. *J. Chem. Soc.*, (B): 428 (1968).
3. Davis, A. and J. H. Golden. *J. Macromol. Sci. Rev. Macromol. Chem.*, C3:49 (1969).
4. Mullen, P. A. and N. Z. Searle. *J. Appl. Poly. Sci.*, 14:765 (1970).
5. Humphrey, J. S., Jr. and R. S. Roller. *Mol. Photochem.*, 3:35 (1971).
6. Humphrey, J. S., Jr., A. R. Shultz and D. B. G. Jacquiss. *Macromolecules*, 6:305 (1973).
7. Gupta, A., A. Rembaum and J. Moacanin. *Macromolecules*, 11:1285 (1978).
8. Ong, E. and H. E. Bair. *ACS Polymer Preprints*, 20:945 (1979).
9. Factor, A. and M. L. Chu. *Polym. Degrad. Stab.*, 2:203 (1980).
10. Clark, D. T. and H. S. Munro. *Polym. Degrad. Stab.*, 4: 441 (1982).
11. Clark, D. T. and H. S. Munro. *Polym. Degrad. Stab.*, 8: 195 (1984).
12. Munro, H. S. and R. S. Allaker. *Polym. Degrad. Stab.*, 11: 349 (1985).
13. Rivaton, A., D. Sallet and J. Lemaire. *Polymer Photochemistry*, 3:463 (1983).
14. Rivaton, A., D. Sallet and J. Lemaire. *Poly. Deg. and Stab.*, 14:1 (1986).
15. Pryde, C. A. "Photoaging of Polycarbonate: Effect of Selected Variables on Degradative Pathways," in *Polymer Stabilization and Degradation*. P. P. Klemchuk, ed. ACS Symposium 280, Chapt. 23 (1985).
16. Webb, J. D. and A. W. Czanderna. *Macromolecules*, 19: 2810 (1986).
17. (a) Factor, A., J. C. Lynch and F. H. Greenberg. *J. of Polymer Sci.: Pt. A: Polymer Chemistry*, 25:3413 (1987). (b) Factor, A., W. V. Ligon and R. J. May. *Macromolecules*, 20:2461 (1987).
18. Schnell, H. *Chemistry and Physics of Polycarbonates*. Interscience Publishers, pp. 99 and 100 (1964).
19. Mark, V. and C. V. Hedges. U. S. Patent 4,560,808 (December 24, 1985).
20. Parris, C. L., R. Dowbenko, R. V. Smith, N. A. Jacobson, J. W. Pearce and R. M. Christenson. *J. Org. Chem.*, 27:455 (1962).
21. Henricks, P. M., M. Linder, J. M. Hewitt, D. Massa and H. V. Isaacson. *Macromolecules*, 17:2412 (1984).
22. O'Gara, J. F., A. A. Jones, C.-C. Hung and P. T. Inglefield. *Macromolecules*, 18:1117 (1985).
23. a) Ligon, W. V., Jr. and R. J. May. *J. Chromatogr.*, 294: 77 (1984); b) Ligon, W. V., Jr. and R. J. May. *J. Chromatogr. Sci.*, 24:2 (1986).
24. Humphrey, J. S., Jr., A. R. Shultz and D. B. G. Jaquiss. *Macromolecules*, 6:305 (1973).
25. Factor, A. *J. of Polymer Sci.*, Part A-1, 7:363 (1969).

T. HJERTBERG[1]
E. MARTINSSON[1]

Influence of Labile Structures and HCl on the Degradation Behaviour of PVC

ABSTRACT

The degradation behaviour of PVC samples with widely different stabilities has been studied, i.e., the rate of dehydrochlorination at 190°C in nitrogen covered the range (6–110) · 10^{-3}% per minute. With decreasing monomer pressure in the polymerization, the stability decreased. These polymers did also show increased content of labile structures such as tertiary chlorine associated with ethyl, butyl and long chain branches as well as internal allylic chlorine. The rate of dehydrochlorination is related to the total content of these structures. In ordinary PVC, tertiary chlorine should be the most important. As expected PVC with decreased content of labile structures showed reduced rate of dehydrochlorination. This result was also accompanied with reduced discolouration, i.e., shorter polyene sequences. Degradations in atmospheres containing HCl proved that this was an effect of decreased content of free HCl due to the increased stability. The experiments further showed that HCl catalyses both random initiation and the propagation of the polyene sequences, i.e. the zipper reaction. We have suggested a mechanism based on ion-pair intermediates to explain the catalytic effect on the polyene elongation.

KEY WORDS

PVC, dehydrochlorination, labile structures, stability, polyene sequences, HCl catalysis.

INTRODUCTION

At temperatures used for processing of PVC the polymer undergoes degradation by dehydrochlorination. The double bonds formed are conjugated in polyene sequences leading to unacceptable discolouration at very low degrees of dehydrochlorination. The polyenes are further active in secondary reactions which causes crosslinking and the formation of aromatic pyrolyzates. The dehydrochlorination of PVC has been observed to occur at temperatures just above the glass transition [1], which is much lower than could be expected. The reason for the low stability has been the subject of many investigations. See References [2–4].

Although different views have been brought up, it has generally been assumed that the low stability of PVC is caused by irregular structures in the polymer activating the carbon-chlorine bond. Based on experiments with low molecular weight model compounds it has been considered that chlorine adjacent to internal double bonds (internal allylic chlorine) and chlorine on branch carbons (tertiary chlorine) are the most labile structures in PVC, and the main reasons for the low thermal stability. The deleterious effect of such structures has also been observed in polymers with increased content of defects obtained by copolymerization [5–7] or chemical treatment [8]. Other possible reasons for the low stability have however recently been suggested as well, e.g., catalysis by ketoallylic groups [9,10] or dehydrochlorination from normal monomer units activated by certain conformations [11,12].

Earlier attempts to reveal the nature of the labile structures actually present in PVC and to relate them to the stability were rather doubtful. By using polymers prepared at reduced monomer pressure we have recently been able to correlate the rate of dehydrochlorination to the amount of tertiary chlorine (mainly associated with butyl branches but also with ethyl and long chain branches) and internal allylic chlorine [13–15]. According to our results, the higher content of tertiary chlorine (1–1.5 per 1000 monomer units) compared to internal allylic chlorine (0.1–0.2 per 1000 monomer units) implies that tertiary chlorine is the most important defect even in ordinary PVC. In accordance with other published data [2,8,16,17] we also found evidence that random dehydrochlorination will occur at the temperatures in question.

In our continued work we have studied in what way the degradation is influenced by a decreased content of labile structures [18,19]. There are several possible ways to obtain fewer defects than normally observed, e.g., by anionic polymerization [20] or by an appropriate treatment

[1]Department of Polymer Technology, Chalmers University of Technology, S-412 96 Göteborg, Sweden.

aiming to substitute labile chlorine with more stable groups. See, e.g. [21 and 22]. As expected, the rate of dehydrochlorination of such samples is lower than that observed for normal PVC. Furthermore is is obvious that PVC with increased heat stability becomes less discoloured as well [18,19,21,23], i.e., the polyene sequences should be shorter. Although completely different techniques have been used to increase the stability, there obviously is a common factor, i.e., the decreased rate of dehydrochlorination. This might lead to a lower concentration of free HCl within the polymer during degradation. We [18,19] as well as Shapiro et al. [23] have therefore suggested that the difference observed between normal and stable PVC regarding degradation behaviour can be related to HCl-catalysis.

The eventual catalytic effect of HCl on the dehydrochlorination has been a controversial point in research concerning the degradation of PVC–see, e.g. [4]. We have shown earlier that the total rate is indeed increased by the presence of HCl [24]. We suggested that HCl may catalyse both the propagation of the polyenes and the random dehydrochlorination. The latter effect was estimated to be the most important and to be responsible for the autocatalytic behaviour often observed.

The present paper summarizes our present view concerning the formation and influence of labile chlorines in PVC. The difference in degradation behavior observed between normal and stable PVC is, however, the major topic. This has given us new possibilities to study the effect of HCl and to clearly demonstrate the catalytic influence of HCl.

EXPERIMENTAL

All details about polymer samples and experimental procedures have been given earlier. The PVC samples include polymers obtained at reduced monomer pressure, relative pressure 0.53–0.97 and polymerization temperature 45–80°C [15,25] as well as ordinary suspension PVC and polymers with increased heat stability. To characterize the structure a number of different techniques have been used: GPC/viscometry to determine the molecular weight distribution (MWD), ^{13}C-NMR analyses after reduction to the corresponding hydrocarbon to determine the branching structure [13], and detection of the changes in $\overline{M}_n$ due to ozone treatment to determine the number of internal double bonds or polyene sequences [14]. The thermal stability has been measured by determining the rate of dehydrochlorination at 190°C. We used either pure nitrogen or nitrogen with 15% HCl as atmosphere and detected evolved HCl by conductivity or thermogravimetry, respectively [19]. We followed changes in the MWD and number of polyene sequences and used UV-visible spectroscopy to qualitatively monitor the polyene sequence distribution.

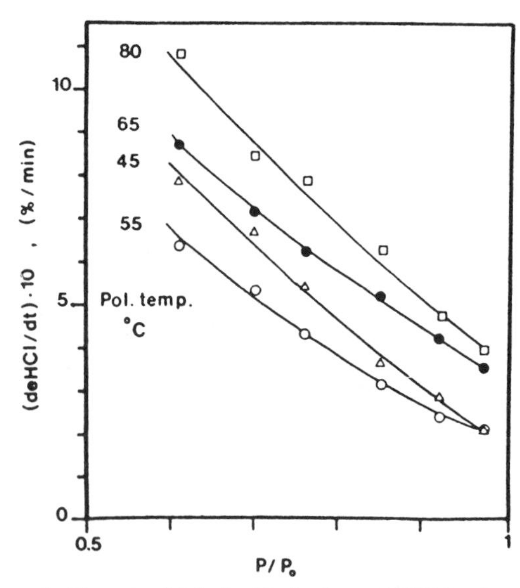

Figure 1. The rate of dehydrochlorination at 190°C in nitrogen as a function of relative monomer pressure (P/P_0) and polymerization temperature.

RESULTS AND DISCUSSION

Earlier attempts to relate structure and stability of PVC samples were hampered by the lack of suitable analytical techniques. The opportunity of using ^{13}C-NMR greatly improved the situation. Besides a low content of defects, e.g., about 1–2 branches with tertiary chlorine per 1000 monomer units, normal commercial PVC should furthermore not show large variations in labile structures. This assumption is based on the fact that the rate of dehydrochlorination normally is found in a rather narrow range, e.g., values between 10–25 · 10^{-3}% per minute were found for 11 commercial samples [26].

In polymerizations at reduced monomer pressure the concentration of monomer in the polymer-monomer gel is less than that at saturated conditions, which normally prevails up to about 70% conversion in ordinary batch polymerizations of vinyl chloride. The decreased monomer concentration should favour other reactions than the propagation step and, consequently, decreased stability should be expected. We have indeed found increased rates of dehydrochlorination for polymers prepared at reduced monomer pressure [13–15,25]. Figure 1 summarizes the effect of both relative monomer pressure and polymerization temperature on the stability.

As expected decreased monomer concentration, i.e., low P/P_0 values, causes decreased stability. Starting from high temperatures a decrease in temperature leads to an increased stability. This trend is also expected as the tendency to side reactions leading to defects should decrease. Figure 1 indicates, however, that there would be an optimum around 55°C. Fortunately, this temperature level coincides with the temperatures most often used in commercial production of PVC.

Even if P/P_0 can be used as a relative measure, the monomer concentration in the polymer gel would be more adequate. For a given value of P/P_0 it turns out that the monomer concentration increases with the temperature [27]. This increase will tend to counteract the expected faster formation of defects due to the effect of the temperature itself, which at least qualitatively is in accordance with the observed optimum. In a comparison based on monomer concentration the 45 and 55°C polymers are quite similar. See Figure 2.

The decreased stability should be accompanied by an increased content of labile structures and it would thus be much easier to detect and quantify them. The content of internal double bonds increases from 0.1–0.3 per 1000 monomer units at $P/P_0 = 0.97$ to about 0.7 at the lowest pressure used. These polymers greatly facilitated the detection of branches, e.g., butyl and long chain branches [28] and later ethyl branches as well [15]. Our own work [13] as well as that of Starnes and coworkers [29–31] have shown that these branches are associated with tertiary chlorine. The long chain branches may, however, partly contain tertiary hydrogen [13]. The content of these structures increases with decreased monomer concentration and increased temperature. The former parameter is the most dominant one and affects the content of internal double bonds in the same way. With increased temperature, however, we could observe a weak tendency to decreased content of internal double bonds. A detailed discussion of analysis methods, influence of polymerization conditions, and mechanisms for the formation of these defects can be found in Reference [15].

The relation between the amount of labile structures and the stability is the most important topic for the present discussion. In an earlier investigation [13,14] involving 55°C polymers only, we concluded that the activity of internal allylic and tertiary chlorine were of the same order. The total sum of ethyl, butyl and long chain branches and internal double bonds can thus be taken as a measure of the content of labile chlorine. As seen in Figure 3 all samples, independent of polymerization temperature, follow the same line. In our earlier work [13,14] the relative content of tertiary and internal allylic chlorine was about constant for most samples, which naturally makes a differentiation more difficult. Our opinion has therefore been criticized [32]. In the present material the relation between the content of these structures varies in the range 2–15 and a plot of the dehydrochlorination rate versus tertiary chlorine comes out similar to Figure 3, while a corresponding one involving internal allylic chlorine does not show any correlation at all. We are very positive about our opinion that tertiary and internal allylic chlorine have about the same reactivity and that their relative importance on the initial dehydrochlorination rate is given by the amount of respective structure. An extrapolation of our data [15] to $P/P_0 = 1$ allows an estimation of the concentration of these structures in ordinary PVC obtained in the same temperature range—see Table 1. At all temperature levels tertiary chlorine is the

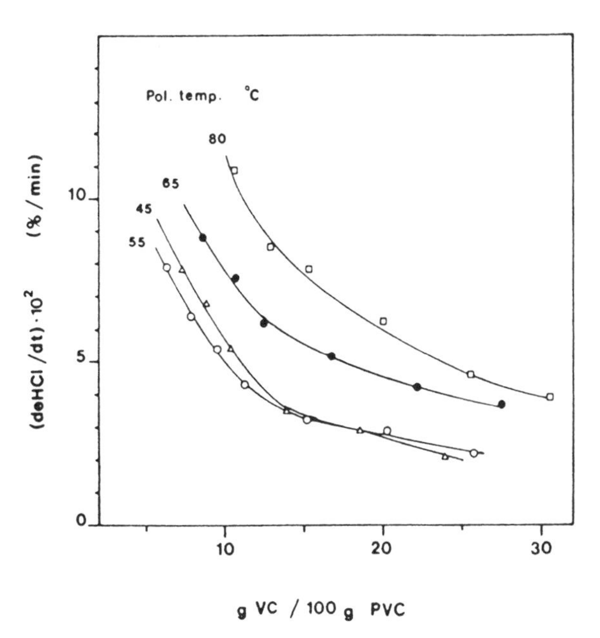

Figure 2. The rate of dehydrochlorination as a function of monomer concentration and polymerization temperature.

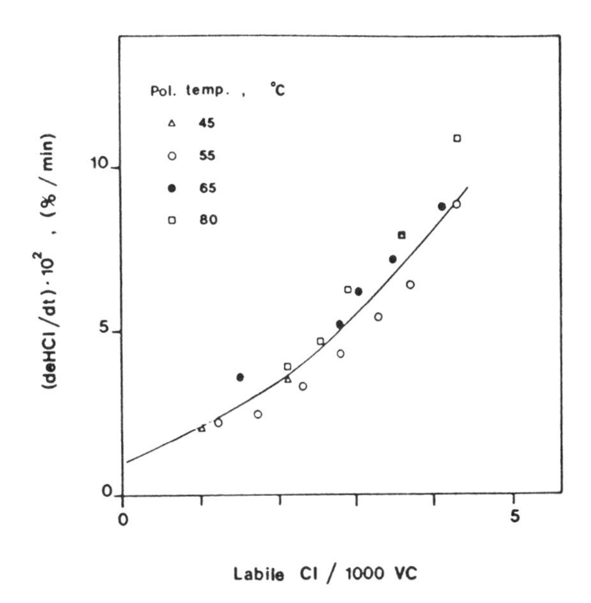

Figure 3. The relation between rate of dehydrochlorination and the content of labile chlorine.

Table 1. Estimated content of labile chlorine in ordinary PVC.

Structure per 1000 VC	Polymerization Temperature, °C			
	45	55	65	80
tertiary	0.7	1.0	1.2	1.8
internal allylic	0.3	0.25	0.2	0.1

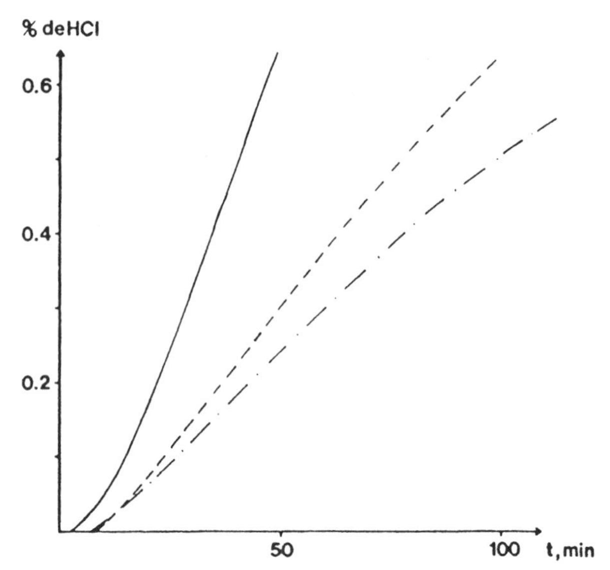

Figure 4. Dehydrochlorination experiments in nitrogen at 190°C with samples A (———); B (----); and C (–·–).

most abundant and its dominance increases with the temperature.

Figure 3 further implies that the relation between the rate of dehydrochlorination and the amount of labile chlorine is non-linear. Considering the discussion below on the effect of HCl this might be an effect of increased stationary concentration of free HCl with increased instability.

An extrapolation to zero content of defects in Figure 3 gives an indication of the importance of random dehydrochlorination. The ultimate stability should thus correspond to about $3 \cdot 10^{-3}\%$ per minute in dehydrochlorination rate with the testing conditions we have used. We have for some time investigated the degradation behaviour of PVC with increased stability compared to that normally observed. Figure 4 shows the course of dehydrochlorination experiments for three samples: samples A and B were obtained by suspension polymerization but sample B has also passed a process aimed to deactivate labile chlorine, and sample C, kindly supplied by AKZO, was obtained by an anionic polymerization. The rates of dehydrochlorination are 17, 7 and $5.5 \cdot 10^{-3}\%$ per minute, respectively. Similar improvements due to different substitutions have been reported earlier [20–22]. It may be mentioned that we have measured a rate of ca. $3 \cdot 10^{-3}\%$ per minute for another sample of anionic PVC, kindly supplied by Dr. Kolinsky. In the case of anionic PVC the high stability can reasonably be referred to a more ideal polymerization mechanism giving many fewer labile structures.

It is very important to note that, independent of the technique used to obtain improved stability, stability is accompanied by a distinctly decreased discolouration at any given level of dehydrochlorination. This effect is illustrated in Figure 5 for samples A and B degraded to 0.2 and 0.4% dehydrochlorination. The improved sample shows much less absorbance at long wavelengths indicating that the polyene sequences are shorter. Sample C is equivalent to B in this respect. By following the formation of new polyenes by ozonolysis we have found that the initiation rate is lower for sample B compared to sample A [18]. This result could be expected after removal of labile structures. Compared at the same level of dehydrochlorination the number of polyenes is higher in sample B, i.e., the polyenes are shorter which is in accordance with the absorbance spectra. We estimated that decreased initiation is of equal importance compared to shorter sequences to explain the lower rate of dehydrochlorination in sample B.

Shorter polyenes have been reported earlier in several instances in connection with degradation of PVC with enhanced stability [21,23,33]. The structures should

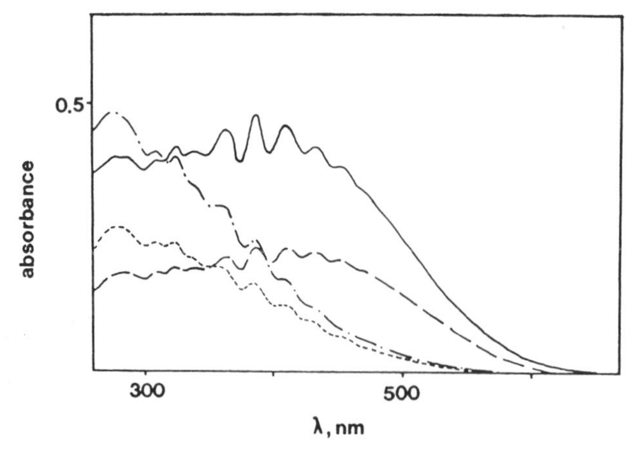

Figure 5. UV-visible spectra of samples A and B degraded to 0.2 and 0.4% dehydrochlorination; (– –), A, 0.2%; (———), A, 0.4%; (----), B, 0.2%; (–·–), B, 0.4%.

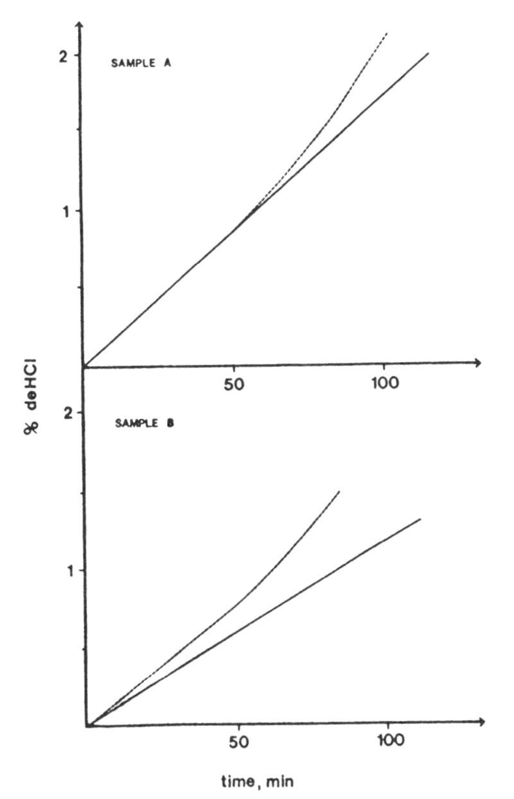

Figure 6. Dehydrochlorination experiments of samples A and B: (———), N_2; (----), N_2 with 15% HCl.

definitely be different, but the low rate of dehydrochlorination is an obvious factor in common for all these polymers. We [18] as well as Shapiro et al. [23] have suggested that the decreased discolouration observed for PVC with improved stability can be related to a lower stationary concentration of free HCl within the polymer during degradation.

A comparison of the dehydrochlorination rate of normal and improved PVC in nitrogen and nitrogen containing deliberately added HCl clearly demonstrated the catalytic effect of HCl. The normal PVC, sample A, was not affected at first, but for sample B the rate was higher from the beginning in the atmosphere containing HCl–see Figure 6. Both samples showed an autocatalytic behaviour at higher levels of dehydrochlorination in the presence of HCl. Ozonolysis showed that this was parallelled by an increased rate of polyene formation. Obviously, random dehydrochlorination is catalyzed by HCl in the presence of polyenes.

The influence of HCl on the initial rate of dehydrochlorination in the case of sample B is a result of longer polyene sequences–see Figure 7. For the normal PVC no difference could be observed in the polyene sequence distribution in accordance with the non-effect on the total rate.

As discussed in more detail in our previous paper [18], there are several possibilities to explain the observed effect of HCl on the polyene length. Inhibition of secondary reactions breaking the conjugation is one, e.g., intramolecular cyclization. The ozonolysis results mentioned above clearly demonstrated, however, that the number of HCl molecules lost in each sequence is lower for sample B, i.e., the sequences should be shorter even without an eventual increase in secondary reactions. The two most probable explanations are HCl-catalysis of the polyene propagation, i.e., the zipper reaction, or HCl-inhibition of a reaction halting the growth of the sequences [18,23]. An example of the latter explanation is cyclization at the growing end after the formation of a trans-cis-trans segment:

Cl Cl Cl $\xrightarrow{-2HCl}$ Cl $\Longrightarrow$ Cl

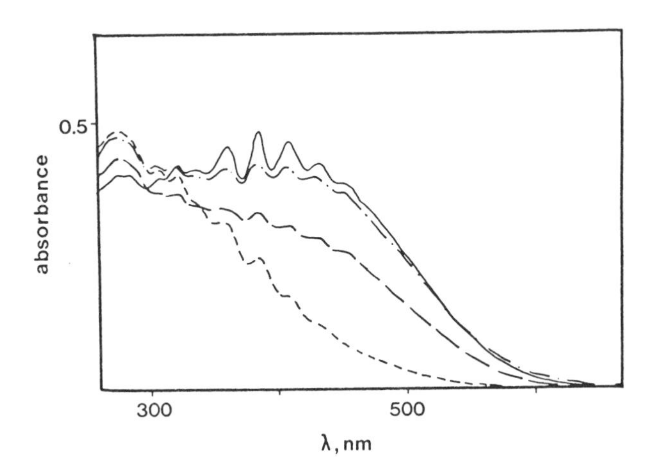

Figure 7. UV-visible spectra of samples A and B degraded to 0.4% dehydrochlorination: (———), A, N_2; (–·–), A, N_2/HCl; (----), B, N_2; (– –), B, N_2/HCl.

It has been suggested that acids may prevent the formation of cis double bonds or catalyze their isomerization into the trans configuration [34]. In such a case, Reaction (1) should be less frequent at higher concentrations of HCl, e.g., in a fast decomposing PVC sample, leading to longer sequences.

We have been able to positively differentiate between the two possibilities by performing a degradation experiment of sample B starting with pure nitrogen as the atmosphere [19]. After a certain level of dehydrochlorination, 0.4%, we added HCl to the carrier gas which led to an increased rate of dehydrochlorination. This should

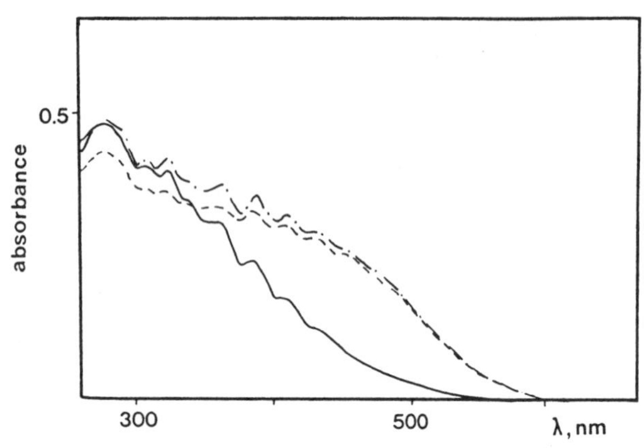

Figure 8. Absorbance spectra of sample B degraded to 0.4% dehydrochlorination; (———), N_2; (— —), N_2/HCl; (—·—), N_2 with addition of HCl during the last two minutes.

correspond to the increased initial rate of dehydrochlorination discussed above (Figure 6), and consequently the new sequences were expected to be longer. Almost immediately, however, the absorbance spectrum shifted to longer wavelengths and become almost equal to that obtained if HCl had been present from the beginning—see Figure 8. The length of all sequences most obviously has been increased. Therefore, Reaction (1) is less likely to be important in determining the sequence length.

Our results, instead, indicate that the polyene sequence length is determined by some equilibrium reaction involving HCl. We have suggested [19] that our observations can be explained on the basis of the ion-pair mechanism for the polyene propagation suggested by Starnes and coworkers [2,34].

H H
Cl Cl Cl
HCl
H H
+ Cl⁻ Cl Cl
HCl
H H
+ Cl δ− Cl⁻ Cl Cl
H δ+
H
Cl Cl

The mere presence of the polar HCl might shift the first equilibrium in favour of the ion-pair, thus increasing the possibility of the next proton abstraction by a chloride ion. Molecular orbital calculations [35,36] have further shown that the positive charge has a strong tendency to concentrate in the centre of the polyenyl cation. This will decrease the ability of the chloride counterion to migrate to the sequence end in order to abstract a proton from the methylene group. Of course, free HCl might associate with the cation thus forcing the chloride ion away from the centre of the polyenyl cation. This would lead to a higher capability to abstract protons and, consequently, to longer sequences. The suggested effects of HCl are indicated in the scheme given above.

CONCLUSIONS

By minimizing the amount of labile structures it is possible to increase the thermal stability of PVC relative to that normally observed. The decreased amount of initiation points will in itself lead to a decreased dehydrochlorination rate. This will result in shorter sequences and still lower total rate. From a practical point of view, the indirect effect—decreased discolouration—is much more important. As random dehydrochlorination from normal monomer units does occur, however, it should not be able to obtain a totally stable PVC.

Our results have shown that HCl catalyzes both random dehydrochlorination and polyene propagation in thermal degradation of PVC. The former effect can only be observed at relatively high degrees of dehydrochlorination and should not be too important in connection with degradation due to processing. The catalytic influence on the polyene propagation is on the other hand of utmost importance because it determines the degree of discolouration. If the amount of HCl within the solid polymer can be kept low enough—by an efficient stabilizer system or by a high inherent stability—the discolouration would become much less severe. As the equilibrium controlling the polyene sequence distribution is very susceptible to HCl, accumulation of free HCl would, however, lead to a rapid and substantial discolouration or blackening.

ACKNOWLEDGEMENTS

Financial support from the Swedish Board for Technical Development and Norsk Hydro AS is gratefully acknowledged.

REFERENCES

1. Palma, G. and M. Carenza. *J. Appl. Pol. Sci.*, 14:1737 (1970).
2. Starnes, W. H. in *Developments in Polymer Degradation*,

Vol. 3. N. Grassie, ed. London:Applied Science Publishers, p. 135 (1981).

3. Braun, D. in *Developments in Polymer Degradation, Vol. 3*. N. Grassie, ed. London:Applied Science Publishers, p. 101 (1981).
4. Hjertberg, T. and E. M. Sörvik in *Degradation and Stabilization of PVC*. E. D. Qwen, ed. London:Applied Science Publishers, p. 21 (1984).
5. Buruiana, E. C., G. Robila, E. C. Bezdadea, V. T. Barbinta and A. A. Caraculacu. *Eur. Polym. J.*, 13:159 (1977).
6. Berens, A. R. *Polym. Eng. Sci.*, 14:318 (1974).
7. Ivan, B., J. P. Kennedy, I. Kenda, T. Kelen and F. Tüdos. *J. Macromomol. Sci. Chem.*, A16:1473 (1981).
8. Ivan, B., J. P. Kennedy, T. Kelen, F. Tüdos, T. T. Nagy and B. Turcsanyi. *J. Polym. Sci. Chem. Ed.*, 21:2177 (1983).
9. Svetly, J., R. Lukas, J. Michalcova and M. Kolinský. *Makromol. Chem., Rapid Commun.*, 1:247 (1980).
10. Lukas, R., O. Pardova, J. Michalcova and V. Paleckova. *J. Polym. Sci. Polym. Lett. Ed.*, 23:85 (1985).
11. Martinez, G., J. Millan, M. Bert, A. Michel and A. Guyot. *J. Macromol. Sci. Chem.*, A12:489 (1978).
12. Martinez, G., C. Mijangos and J. Millan. *Eur. Pol. Sci.*, 21:387 (1985).
13. Hjertberg, T. and E. M. Sörvik. *Polymer*, 24:673 (1983).
14. Hjertberg, T. and E. M. Sörvik. *Polymer*, 24:685 (1983).
15. Hjertberg, T. and E. M. Sörvik. *ACS Symposium Series*, 280:259 (1984).
16. Starnes, W. H., I. M. Plitz, D. C. Hische, D. J. Freed, F. C. Schilling and M. L. Schilling. *Macromolecules*, 11: 373 (1978).
17. Starnes, W. H., R. C. Haddon, D. C. Hische, I. M. Plitz, C. L. Schosser, F. C. Schilling and D. J. Freed. *Polym. Prepr.*, 21(2):176 (1980).
18. Hjertberg, T., E. Martinsson and E. Sörvik. *Macromolecules* (in press).
19. Martinsson, E., T. Hjertberg and E. Sörvik. submitted.
20. Jisová, V., M. Kolinský and D. Lim. *J. Polym. Sci.*, A-1 (8):1525 (1970).
21. Mitani, K., T. Ogata, M. Nakatsukasa and Y. Mitzutani. *Polymer*, 21:1463 (1980).
22. a) Starnes, W. H. and I. M. Plitz. *Macromolecules*, 9:633 (1976). b) I. M. Plitz, R. A. Willingham and W. H. Starnes. *Macromolecules*, 10:499 (1977).
23. Shapiro, J. S., W. H. Starnes, I. M. Plitz and D. C. Hische. *Macromolecules*, 19:230 (1986).
24. Hjertberg, T. and E. M. Sörvik. *J. Appl. Pol. Sci.*, 22: 2415 (1978).
25. Hjertberg, T. and E. M. Sörvik. *J. Pol. Sci. Chem. Ed.*, 24:1313 (1986).
26. Abbås, K. B. and E. M. Sörvik. *J. Appl. Polym. Sci.*, 20:2395 (1976).
27. Nilsson, H., C. Silvergren and B. Törnell. *Eur. Polym. J.*, 14:737 (1978).
28. Hjertberg, T. and E. Sörvik. *J. Polym. Sci. Polym. Lett. Ed.*, 19:363 (1981).
29. Starnes, W. H., F. C. Schilling, I. M. Plitz, R. E. Cais, D. J. Freed and F. A. Bovey. paper presented at the *Third Int. Symp. on PVC, August 1980, Cleveland, USA*, preprints, p. 58.
30. Starnes, W. H., F. C. Schilling, I. M. Plitz, R. E. Cais and F. A. Bovey. *Polym. Bull.*, 4:555 (1981).
31. Starnes, W. H., F. C. Schilling, I. M. Plitz, R. E. Cais, D. J. Freed, R. L. Hartless and F. A. Bovey. *Macromolecules*, 16:790 (1983).
32. Braun, D. B. Böhringer, F. Tüdos, T. Kelen and T. T. Nagy. *Eur. Poly. J.*, 20:799 (1984).
33. Suzuki, T. *Pure Appl. Chem.*, 49:539 (1977).
34. Starnes, W. H. and D. Edelson. *Macromolecules*, 12:797 (1979).
35. Haddon, R. C. and W. H. Starnes. *Polym. Prepr.*, 18:505 (1977).
36. Haddon, R. C. and W. H. Starnes. *ACS Adv. Che. Ser.*, 169:333 (1978).

P. HRDLOVIČ[1]
I. LUKÁČ[1]

Light Degradation of Ketone Polymers

ABSTRACT

Specific features of the photochemical behaviour of the carbonyl group in polyketones are discussed. Attention is focused on polyketones with γ-hydrogen with respect to the carbonyl group, which exhibit the Norrish type II reaction. Absorption, emission and transient spectra are correlated with quantum yields and Stern-Volmer constants for poly/1-/4-X-phenyl/-2-propen-1-ones where X is various substituents—alkyl, halogen, methoxy, phenyl, etc. The quantum yields of the Norrish type II reaction (β-photoelimination) are comparable to those observed for low molecular mass analogues between 0.2 and 0.4. The macromolecular character of the substrate is observed in the emission spectra and in quenching. The red shift of the emission of the homopolymers compared to the model compound is due to the defects in the polymer chain. The rate constants for quenching measured directly by flash photolysis or estimated from Stern-Volmer constants for low molecular mass quenchers as conjugated dienes, naphthalene, biphenyl follow the order k_q/model $>$ k_q/copolymer $\geq$ k_q/homopolymer. Polymer-bound quenchers exhibit the same efficiency as the free ones, when local concentration effects are taken into consideration.

INTRODUCTION

The carbonyl group is the simplest chromophore which absorbs in the near UV range around 300 nm and exhibits various photochemical reactions [1,2]. The type and the mechanism of the photochemical reaction is determined by the structure of the surroundings of the carbonyl group and by the milieu as well. In application and processing of polymers, carbonyl groups were found long ago to take part in thermal and photochemical oxidation of carbonaceous polymers [3]. For this reason, great attention was devoted to the photochemical behaviour of the carbonyl group in polymers. The role of the carbonyl group as initiator, propagator and end product during photooxidation and photolysis of the carbonaceous polymers is still the subject of discussion [3].

[1]Polymer Institute, Centre for Chemical Research, Slovak Academy of Sciences, CS-842 36 Bratislava, Dubravska cesta 10, Czechoslovakia.

Photochemical behaviour of the carbonyl group in simple and complex molecules have been investigated intensively in gas and liquid phases [1,2]. The application of the results is limited for the estimation of photochemical behaviour of this group in solid polymers. The reasons are:

1. In macromolecules the chromophores are bound together even in diluted solutions. Consequently, there can be stronger interaction of chromophores than in low molecular mass compounds. Some isolation of the chromophore in a macromolecule can be achieved via copolymerization with a photochemically inert comonomer.
2. Photochemical degradation of polymers as construction materials occurs in the solid phase [3]. On the contrary, photochemical reactions are investigated in solution or in gas [1,2].
3. Ageing of polymers, where carbonyl groups take part, occurs in the presence of oxygen. Photochemistry of carbonyl groups originates mostly from the triplet state [1,2]. It is well known that oxygen quenches the triplet state efficiently forming singlet oxygen which exhibits its own photochemistry. Moreover, oxygen with its addition reactions competes with the recombination of radicals giving hydroperoxides and peroxides, which again, are photochemically active.

Irradiation of the carbonyl group in a low molecular mass compound in diluted solution leads to an unequivocal photochemical reaction [1,2]. The same chromophore present in the solid phase in an oxygen atmosphere, after irradiation causes a complex sequence of photochemical oxidation and fragmentation reactions [3].

Typical photochemical reactions of the carbonyl group include [1,2]: splitting of the carbon-carbon bond in the neighbourhood of carbonyl/α-splitting/ and intra- and intermolecular hydrogen abstraction. Hydrogen in γ-carbon with respect to the carbonyl group is the most easily abstracted one in the intramolecular hydrogen abstraction. As a result the carbon-carbon β-bond with respect

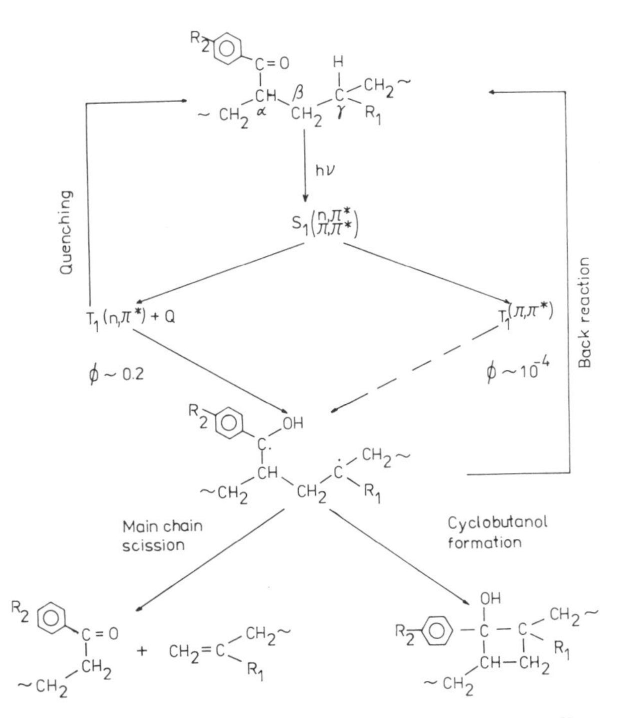

Figure 1. Reaction scheme of main chain scission (Norrish type II reaction) and formation of cyclobutanol.

to carbonyl is broken (Norrish type II reaction). A decrease of the molecular mass brings about deterioration of the mechanical properties [3]. Both photochemical reactions of carbonyl in a polymer, α-splitting and β-photoelimination, result in a decrease of the molecular mass when this chromophore is bound to the polymer or is formed during oxidation. Since the first studies on photolysis of poly/1-buten-3-one/ in solution [4] and in film [5], many investigations on the photochemistry of the carbonyl group in polymers have been made. Different aspects of photolysis and photooxidation of polyketones were followed, such as product analysis, influence of quenchers and inhibitors, influence of solvents and polymer matrix, energy transfer and sensibilization properties [6,7].

In the review article we summarize the results from photochemical studies of the carbonyl group in the side chain adjacent to the main chain according to Figure 1, where R_1 is the same chromophore as that entering the reaction for homopolymers (or another group depending on the comonomer) and R_2 are various substituents, such as alkyl, halogen, methoxy, phenyl, etc. The dominant reaction is the Norrish type II reaction when the γ-carbon has hydrogen as a substituent. The goal of this paper is to correlate the spectral and photochemical data in order to point out the specific features of the macromolecular structure and to evaluate at least partially the role of this chromophore in the complex process of ageing.

SYNTHESIS OF MONOMERS AND POLYMERS

Preparation and properties of the most common ketone monomers have been described in a review article [8]. There are several ways for the preparation of derivatives of 1-phenyl-2-propen-1-one reported in the literature [6,8]. The most frequently used method is the Friedel-Crafts acylation of aromatic hydrocarbons with chloride of 3-chloro-propanoic acid [6]. An alternative route is the synthesis through Mannich salts. It can be used for the preparation of monomers having phenyl substituted with electron donor and electron acceptor substituents. Mannich salts are easy to prepare and purify.

Monomers with a vinyl group adjacent to a carbonyl are very reactive and easily polymerize even in the absence of an initiator of radical polymerization. A polymer with the higher content of isotactic dyads was prepared by the polymerization of 1-phenyl-2-propene-1-one with dialkylzinc [9]. With the same aim lithium tert.-butoxide was used in the polymerization of 1-/4-*X*-phenyl/-2-propen-1-one [10]. The content of isotactic dyads increased.

Copolymerization of vinyl ketones with other vinyl monomers provides further possibilities to modify the photochemical behaviour of the original chromophore with the change of the substituent on the γ-carbon. The choice of suitable comonomers can be made on the basis of the copolymerization parameters available for the limited number of monomer pairs, the second monomer is usually styrene. The dependence of the copolymerization parameters on benzene ring substituents was established for 1-/4-*X*-phenyl/-2-propen-1-one (M_1) where X is 4-OCH_3, 4-CH_3, H, 4-F, 4-Cl, 4-$COCH_3$, 3-Cl, 3-CH_3 and 3-CN with styrene (M_2) while r_1 varied between 0.15 and 1.15 and r_2 between 0.04 and 0.19 [11]. For $X = $ H (M_1) and styrene (M_2) the copolymerization parameters $r_1 = 0.48$ and $r_2 = 0.21$ [12] are in good agreement with those reported [11]. For $X = $ H (M_1) copolymerization parameters with vinyl acetate (M_2) in benzene are $r_1 = 6.5$ and $r_2 = 0.1$ [13]. For $X = $ H (M_1) and butyl acrylate (M_2) r_1 and r_2 are 1.15 and 0.74 respectively [14]. For $X = $ 2-methoxy-5-tert.-butyl (M_1) and butyl acrylate (M_2) in benzene solution r_1 and r_2 are 3.27 and 0.21 [15]. Copolymerization parameters for 1-buten-3-one (M_1) with common vinyl monomers (M_2) are $r_1 = 1.6 - 8.3$ and $r_2 = 0.05 - 0.65$ [8]. The copolymerization parameters indicate that copolymer enriched with vinyl ketone is formed at low conversion. With the increasing conversion the content of vinyl ketone decreases. As a result, the products of heterogeneous chemical composition are formed. The majority of photochemical data was obtained on copolymers of heterogeneous chemical composition [6]. Lukáč and co-workers [16] have recently investigated photolysis of carefully prepared copolymers of 1-/4-carbethoxyphenyl/-2-propen-1-one with styrene, vinyl acetate, ethyl acrylate, phenyl acrylate, acrylonitrile and 1-buten-3-one. The authors concluded that some effects on copolymers observed in the past are mainly due to their heterogeneous chemical composition.

SPECTRAL CHARACTERISTICS

The primary goal of spectral measurements is to obtain information about the nature and dynamics of the excited states. The spectral data of low molecular mass compounds are usually measured under conditions in which intermolecular interactions are excluded. Absorption, emission and transient spectra of the low molecular mass compounds are usually measured in diluted or solid solutions. Macromolecules, however, exhibit chromophores mutually tightly bound in the main or side chain. This arrangement brings about some effects (similar for example to crystals) which complicate (in some respects) the interpretation of spectral data. On the other hand, they offer additional information on the system. The specific effects of macromolecules include, e.g., intramolecular energy transfer along the polymer chain, traps or sinks as a result of structural defects, and specific quenching.

Absorption Spectra

Ultraviolet absorption spectrum of a macromolecule as an assembly of chromophores reflects more or less the absorption of the isolated chromophores. As a result the absorption spectra of polyketones [16–26] are similar to those of the low molecular mass aryl alkyl carbonyl which exhibit an intense band above 250 nm ($\epsilon \sim 10^4$ $mol \cdot dm^{-3} \cdot cm^{-1}$) and a less intense band around 330 nm ($\epsilon \sim 10^2$ $mol \cdot dm^{-3} \cdot cm^{-1}$). Some small differences are observed between the absorption spectra of model compounds and polymers which can be caused by several factors. Packing chromophores along the polymer chain can bring about a more strained conformation of chromophores and consequently a decrease of the conjugation in the chromophore. As a result the blue shift and less intense band could appear in the absorption spectrum of the polymer as compared to the model. It seems that simple aryl carbonyl chromophores in 1,3 arrangement along the main chain do not show this effect. On the other hand, polymer bound chromophores can form ground or excited state complexes. Moreover, the microenvironment of tightly bound chromophores in a macromolecule differs from the model compound because of a different degree of solvation. For the majority of pairs of model compound-homopolymer, some differences were observed in the long wavelength region of the absorption spectra. Usually a slight bathochromic shift of the longest wavelength band was observed for homopolymer as compared with the model. This shift is due to a more polar environment in a homopolymer as compared to the

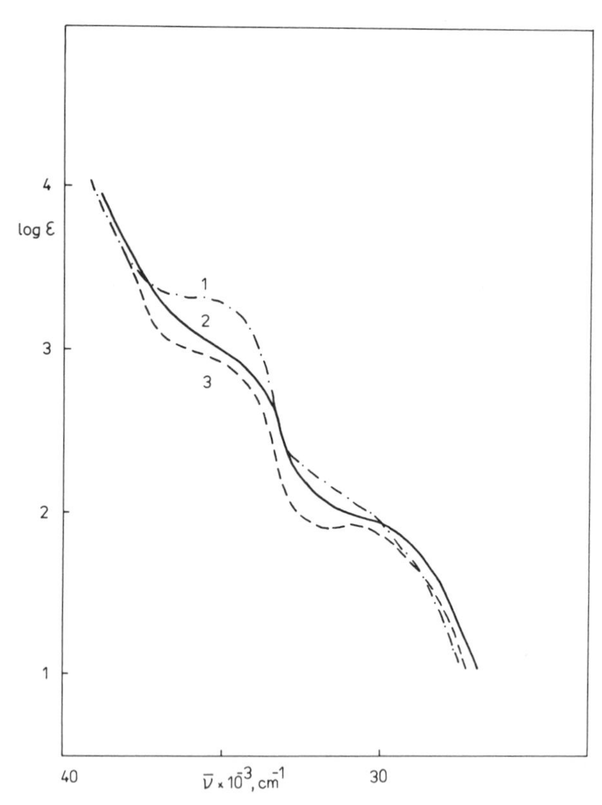

Figure 2. Ultraviolet absorption spectra in chloroform of 1-phenyl-3-chloro-1-propanone (model compound) (1); poly(1-phenyl-2-propen-1-one) prepared in anionic (2); and radical (3) polymerization.

solvated model. Since the changes are slight they can be influenced by the choice of the solvent. In fact, negligible differences were observed in the absorption spectra of homopolymers prepared by radical and anionic polymerization (Figure 2) [10]. Since anionic polymers do not have high molecular mass (maximum 10^4) and the content of isotactic dyads increases only slightly, the question of the influence of the microstructure of the polymer chain on the absorption is to be answered.

A more pronounced effect on the UV spectra is produced by a substitution at position 4 on the benzene ring of aryl alkyl carbonyl. Electron donor substituents strongly influence the π-π^* band and shift it bathochromically (Figure 3) [20]. The less intense n-π^* band becomes a shoulder and disappears. Halogens show a similar effect [23]. The change from phenyl to 2-naphthyl shifts the absorption bathochromically and no n-π^* band can be seen in the UV spectrum [20]. This shift indicates that the lowest singlet excited state has the π-π^* character. These changes in the UV spectra are reflected in the photochemical reactivity.

Emission Spectra

Aryl alkyl carbonyl exhibits emission as one path for relaxation of the excited state. Since this type of chro-

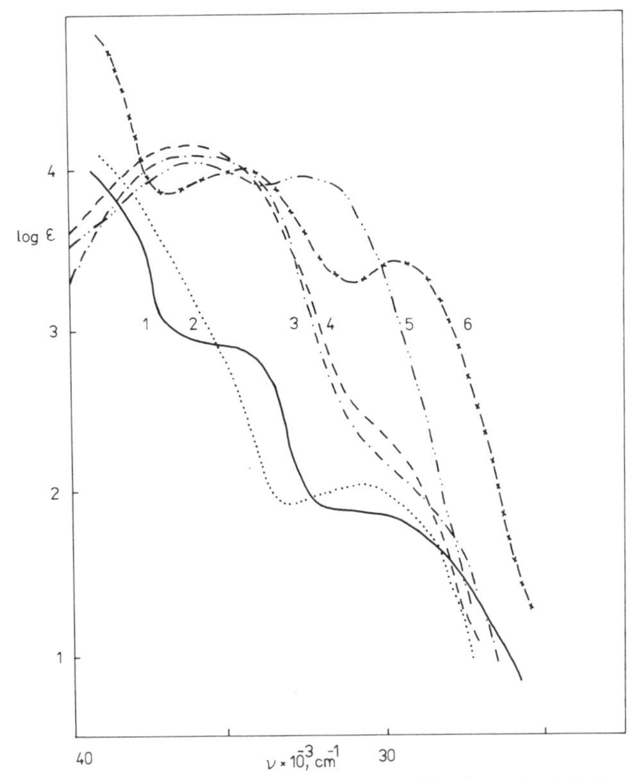

Figure 3. Ultraviolet absorption spectra in chloroform of poly(1-(4-*x*-phenyl)-2-propen-1-one) for x = H (1); F (2); J (3); OCH_3 (4); phenyl (5); and for poly(1-(2-naphthyl)-2-propen-1-one) (6).

mophore does not show any fluorescence, the observed emission is phosphorescence which is very intense at low temperature (77 K) [2,6,7]. Model compounds of the type 1-/4-*X*-phenyl/-3-chloro-1-propanone and non-photoreactive polyketones show phosphorescence at room temperature as well [25,26]. For examining the emission spectra of the low molecular mass compounds at 77 K mixture of solvents is used which secures the formation of a good optical glass. The use of a mixture of solvents like hydrocarbons, ether or alcohol leads to problems with the solubility of polymers. Moreover, oxygen must be removed from these solutions. It is convenient to measure emission spectra of polymers in the form of films 50–100 μm thick dipped in liquid nitrogen with no interference of oxygen [6]. Poly(methyl methacrylate) was used as a matrix for model compounds and polymers which do not yield a self-supporting film. In the case of homopolymers doped in poly(methyl methacrylate), films were opaque indicating low compatibility with the respective polymers. The low optical quality of the film does not influence the emission spectrum. This was demonstrated for several homopolymers. The typical emission spectrum of the model compound 1-/4-fluorophenyl/-3-chloro-1-propanone, the corresponding homopolymer and the copolymer with styrene (Figure 4) are well resolved vibrationally like unsubstituted derivatives which indicates that the fluoro substituent exerts a small influence on the emission. Although the vibrational

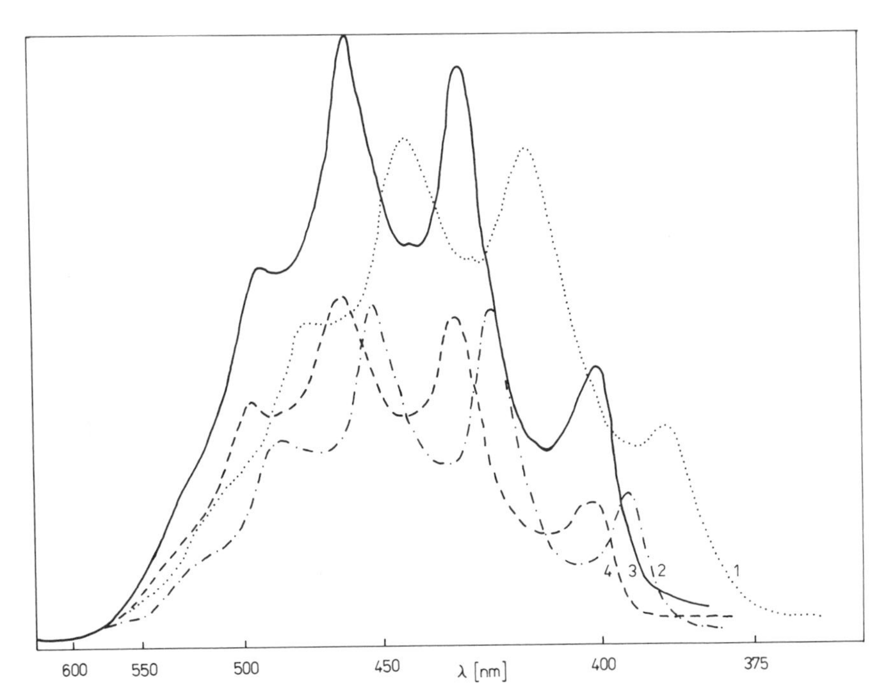

Figure 4. Emission spectrum at 77 K of (1) 1-(4-fluorophenyl)-3-chloro-1-propanone doped in poly (methyl methacrylate) film (5 wt./wt.%); (2) poly(1-(4-fluorophenyl)-2-propen-1-one-CO-styrene) film; (3) poly(1-(4-fluorophenyl)-2-propen-1-one) doped in poly(methyl methacrylate) film (5 wt./wt.%); (4) poly(1-(4-fluorophenyl)-2-propen-1-one) film.

structure is well resolved, the stretching vibration of the carbonyl group (1660–1680 cm^{-1}) is not preserved in the whole region especially on the long wavelength side. This lack of stretching is due to the polarity of poly (methyl methacrylate) as a matrix [27]. Better vibrational resolution of the carbonyl compounds can be reached in a non-polar matrix, for example polyethylene or polypropylene [27]. Some technical problems–high temperature of melting and incompatibility–hinder the preparation of polypropylene film doped by homopolymer. For this reason only the emission spectra of the model compounds of the type 1-/4-*X*-phenyl/-3-chloro-1-propanone in polypropylene and poly(methyl methacrylate) were compared [6]. In spite of the fact that polypropylene is non-polar and microcrystalline, no large shifts were observed in the emission spectra of the model compounds. Contrary to polypropylene the emission of cyclic ketones like xanthone or anthrone undergoes a great change in going from polypropylene to poly(methyl methacrylate) matrix [27]. No unequivocal shift of the O–O band was observed for the model compounds, which are in fact acyclic ketones, in going from polypropylene to poly(methyl methacrylate) matrix. For the low molecular mass aryl alkyl ketones with the lowest n-π^* triplet only a slight, hypsochromic shift is observed in going from polypropylene to poly(methyl methacrylate) matrix. This is an opposite shift to the one observed for anthrone and xanthone [27]. Aryl alkyl ketones, ring-substituted with electron donating substituents, show a slight bathochromic shift in going from non-polar to a polar matrix.

A pronounced bathochromic shift was observed for homopolymers in comparison with the model compounds for nearly all pairs. Table 1 gives some examples. This bathochromic shift of the emission of homopolymers can be related to tight packing of chromophores along the polymer chain. It represents an interaction of 10–20 kJ·mol^{-1}. The red shift of the emission can be an interaction of the side chromophores, one in the excited state and one in the ground state (triplet excimer) or the formation of an energy trap in the main chain due to a structural defect, which is difficult to identify by other methods. There are some indirect evidences supporting the second proposal:

1. Emission spectra of model compounds–1,5-diphenyl-1,5-pentadione and 1,5-diphenyl-2,4-dimethyl-1,5-pentadione are like those of monoketone 1-phenyl-3-chloro-1-propanone [6].
2. Anionic polyketones with the low molecular mass showed a less pronounced shift than the high molecular mass radical polyketones [10].
3. Copolymers with different content of carbonyl group exhibit emission spectra somewhere between those of model and homopolymer depending on composition (Table 1) [22].

Transient Spectra

It is evident from Figure 1 that the Norrish type II reaction yields at least two transients–triplet state and 1,4-biradical–the behaviour of which strongly influences the course of the reaction. Naturally, the low and high molecular mass ketones showing the Norrish type II reaction were the subject of the first time resolved measurements [28,29]. Typical transient spectrum of the model compounds and polyketones shows one maximum [25,26]. Transient absorption of poly(1-/4-fluorophenyl/-2-propen-1-one) and poly(1-/3,4-dimethoxyphenyl/-2-propen-1-one) (Figure 5) belongs to the triplet state which is for 4-fluoro derivative reactive and for 3,4-dimethoxy derivative non-reactive. Methoxy derivatives of ketones exhibit a rather long-living triplet which is easy to follow. Some short-living triplets can overlap with the signal of 1,4-biradical. In order to minimize

Table 1. Emission spectra of copolymers of carbonyl monomers with styrene compared with homopolymers and model compounds in film at 77 K [6].

Copolymer	CO[a] wt./wt.%	λ[b] nm			
poly(1-phenyl-2-propen-1-one)	100	409	437	470	500[s]
poly(1-phenyl-2-propen-1-one-co-styrene)	28.6	400	428	458	486
	19.0	397	424	455	486
	10.2	397	422	451	483
1-phenyl-3-chloro-1-propanone[c]		393	420	450	480[s]
poly(1-(4-methoxyphenyl)-2-propen-1-one)	100	430[s]	475		
poly(1-(4-methoxyphenyl)-2-propen-1-one-co-styrene)	88.8	425[s]	457		
	22.5	410	437	452[s]	
	12.7	410	437	452[s]	
1-(4-methoxyphenyl)-3-chloro-1-propanone)[c]		410	430	453	

[a]Content of carbonyl monomer.
[b]s means shoulder.
[c]Content of 5 wt./wt.% in polymethylmethacrylate matrix.

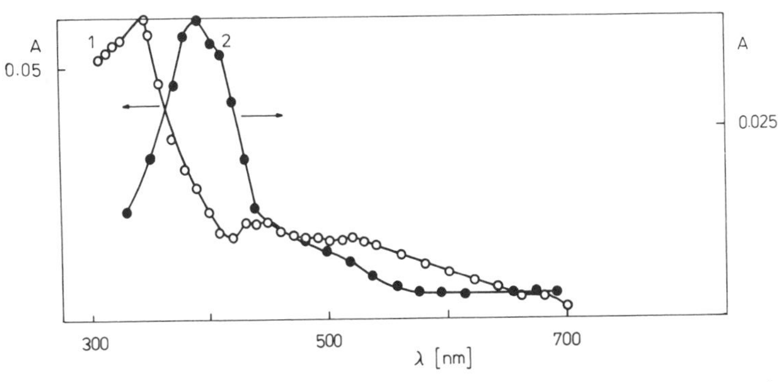

Figure 5. Transition spectrum of poly(l-(4-fluorophenyl)-2-propen-l-one) (1) and poly(l-(3,4-dimethoxyphenyl)-2-propen-l-one); (2) in chloroform at room temperature.

these problems the decay was followed in the interval between 350 and 390 nm. Table 2 summarizes the main characteristics of the transient spectra. Great difference in the lifetime of the model compounds and polymers is due to the fact that the former do not exhibit the Norrish type II reaction while the latter do. A comparison of the transient spectra of the model compounds and homopolymers leads to the conclusion that there is no difference in the position of the maximum. On the other hand, strong electron-donor substituents in position 4 on the benzene ring cause the bathochromic shift. Some model compounds and polymers show additional absorption above 400 nm which decays slower than the main band. Residual absorption observed for all models and polymers was weak. At shorter times it is due to biradical, at longer times it is due to the presence of ketoneols which can be formed by inter- and intramolecular hydrogen abstraction.

In the case of non-reactive 3,4-dimethoxy derivatives the triplet lifetime was longer for polymer than for the model compound. Substitution at position 4 of phenyl ring in polymers systematically increases the triplet lifetime with the exception of fluorine. Methoxy substitution leads to long lifetimes as expected from the π,π^* nature of the low-lying triplet. It was suggested that the 3,4-dimethoxy derivative exhibits efficient self-quenching [30]. Since the transient signal was rather strong for homopolymer and model compound and no dependence on the concentration was observed, the self-quenching does not operate for this chromophore.

On the other hand, biphenyl carbonyl and 2-naphthyl carbonyl chromophores show self-quenching. The rate constant for self-quenching was for biphenyl carbonyl estimated to be around 10^7 $dm^3 \cdot mol^{-1} \cdot s^{-1}$. The self-quenching of 2-naphthyl carbonyl is even more effective [26].

Transient absorption is quenched by typical triplet quenchers, like naphthalene, conjugated dienes, oxygen and N-oxyl. This quenching is additional evidence that the transients originate in triplet state.

PHOTOLYSIS IN SOLUTION AND IN FILM

The main features of the photolysis of poly(1-/4-X-phenyl/2-propen-l-one) with γ-hydrogen that can be abstracted are evident from Figure 1, which agrees with the scheme suggested by Wagner for the low molecular mass compounds [31]. It is characteristic of polyketones that the β-carbon-carbon bond with respect to the carbonyl group is in the main chain. Its breaking brings about main chain scission and a decrease in molecular mass. Various techniques were used for monitoring the decrease of molecular mass: osmometry, viscometry, low-angle light scattering, and GPC. The most frequently used technique is viscometry which is experimentally easily accessible, reproducible and sensitive at the beginning if the molecular mass of the starting polymer is high. For this reason viscometry was used to follow the degradation of ketone-containing polymers [16–26]. The extent of degradation is given by the number of main chain scission calculated according to

$$s = \overline{M}_o/\overline{M}_t - 1$$

where $\overline{M}_o$ and $\overline{M}_t$ are number average molecular masses at the beginning and after irradiation for the time t. This quantity can be calculated with an acceptable accuracy using viscometric data according to

$$s = ([\eta]_o/[\eta]_t)^{1/\alpha} - 1$$

where $[\eta]_o$ and $[\eta]_t$ are the limiting viscosity numbers at the beginning and at the time t, α is the coefficient of the Mark-Houwink equation which relates the molecular mass and the limiting viscosity number.

Quantum yield of the main chain scission is in fact the quantum yield of the Norrish type II reaction if the other routes leading to chemical products are negligible (Figure 1). It can be calculated using equation

$$\phi = c_p \cdot s/\overline{M}_n(o) \cdot I_p \cdot t$$

Table 2. Transient spectra of poly(1-(4-X-phenyl)-2-propen-1-ones) in chloroform at room temperature [25,26].

X	Substrate	λ^{max} nm		τ_T μs	k_q[d] $M^{-1}s^{-1}$	K_{SV}[f] (M^{-1}) Naph.	Biphenyl	Diene
F	Model[a]	347	436[b]	2.576	5.30×10^9			
	Homopolymer	345	449[b]	0.075	1.14×10^9	62	30	
	Copolymer/S	325		0.072	7.15×10^8	190	43	
Cl	Model[a]	350		7.820	3.90×10^9			
	Homopolymer	350		0.200	1.26×10^9	260	48	
Br	Model[a]	381		0.060	5.72×10^9			
	Homopolymer	380		0.107	1.77×10^9	110	30	
	Copolymer/(Styrene)			0.034	1.59×10^9	47	17	
CH_3CO	Model[e]			0.456	4.54×10^9			
	Homopolymer	361		0.489	1.52×10^9	381[g]	4.4[g]	
	Copolymer/(Styrene) (43.2%)			0.459	1.47×10^9	770	3.2	620
	Copolymer/(MMA) (41.8%)	352		6.894	1.96×10^9	9670	65	2400
	Copolymer/(MMA) (6.0%)	352		19.583	1.82×10^9			
$COOC_2H_5$	Homopolymer	349		0.126	5.57×10^8	170		
CH_3	Homopolymer	347		0.288	9.72×10^8	436	133	
C_2H_5	Homopolymer	349		0.293	8.9×10^8	345	78	
$(CH_3)_3C$	Homopolymer	350		0.307	5.7×10^8	336	96	
CH_3O	Model[a]	393		5.800[c]	4.25×10^9			
	Homopolymer	389	(640)[b]	4.600[c]	7.64×10^8	1700[h]	530[h]	
	Copolymer/(Styrene) (22.2%)	391		4.400[c]	1.33×10^9	1000[h]	350	
	Copolymer/(MMA) (22.3%)	398		9.800[c]	1.15×10^9	7100[h]		
$3,4(CH_3O)_2$	Model[a]	393		12.300[c]	2.60×10^9			
	Homopolymer	392		28.300[c]	2.56×10^8			
	Copolymer/(Styrene)	380		20.400[c]	7.16×10^8			
	Copolymer/(MMA)	390		29.900	6.71×10^8			
4 C_6H_5	Model[a,i]	425		140	5.02×10^8	(TEMPO)		
	Homopolymer (anionic)[i]	425		145				
	Homopolymer (radical)[i]	400		136				
	Copolymer/(MMA)[i]	425		255	1.8×10^8	(TEMPO)		

[a]Model is 1-(4-X-phenyl)-3-chlorpropan-1-one.
[b]Weak additional absorption.
[c]Extrapolated to zero laser does.
[d]Quencher was 2,5-dimethyl-2,4-hexadiene or 4-acetyloxy-2,2,6,6-tetramethylpiperidine-N-oxyl (TEMPO).
[e]Model is 1,4-diacetylbenzene.
[f]Stern-Volmer constant determined by quenching of photolysis in benzene.
[g]Dioxane.
[h]Chlorbenzene.
[i]Benzene.

where c_p is the concentration of irradiated polymer solution in g·dm^{-3} or mass of polymer film in g·cm^{-2}, $\overline{M}_n(o)$ is the original number average molecular mass, s is the number of the main chain scission, I_p is the radiation intensity absorbed by polymer given in E · dm^{-3}·s^{-1} for solution or E · cm^{-2}·s^{-1} for film and t is the irradiation time. Assuming the validity of the Lambert-Beer Law, the intensity of radiation absorbed by polymer is given by

$$I_p = I_o(1 - 10^{-\epsilon \cdot c \cdot d})$$

where I_o is the original intensity and the product $\epsilon \cdot c \cdot d$ is the absorbance of the polymer solution or film.

Although viscometry is a simple and rapid method, it has some inherent problems.

First of all, poly (aryl alkyl ketones) are in fact special polymers and are not well characterized in regard to their solution properties. The coefficients α and K, relating limiting viscosity number to molecular mass, were determined for poly /1-phenyl-2-propen-1-one/ using samples prepared by stepwise degradation. The molecular mass was determined osmometrically [32]. The values α and K were checked by GPC [22,33,34]. Low-angle light scattering combined with automatic viscometry has been recently used for determination of α and K of several derivatives of poly(1-/4-X-phenyl/-2-propen-1-one) [35].

The value of quantum yields is strongly influenced by $\overline{M}_n(o)$. The polyketones were prepared by radical polymerization$-\overline{M}_w/\overline{M}_n$ is about 2. This ratio does not substantially change during random degradation; therefore, it is possible to substitute $\overline{M}_n(o)$ by $M_v/2$ if $\overline{M}_w = M_v$ holds. Since values of α and K were often taken from [32], the approximation $\overline{M}_n \doteq M_v$ was used.

The irradiation according to Figure 1 does not lead to a decrease of the absorbing chromophores except for the formation of cyclobutanol. During the reaction, however, the number of chromophores able to react decreases, and the number of unsaturated ketones at chain ends increases. No correction was done for absorption by chromophores which do not undergo photochemical reaction. If the extent of degradation is small (s is about 5 to 10 scissions), then the simple calculation shows that the correction for internal screening is not needed.

The ratio of cyclobutanol formation to β-photoelimination depends on substituents in the γ-position for the low molecular mass compounds [36,37]. If γ-substituents are electron acceptors, the quantum yields of cyclobutanol formation are low or negligible [37]. Consequently, low quantum yields of cyclobutanol formation can be expected for homopolymers.

The formation of photostable cyclobutanol from copolymers of carbonyl monomers with styrene may be expected to be similar to that from 1,4-diphenyl-1-butanone with a quantum yield of about one tenth of main chain scission [37]. This means that for a copolymer with $\overline{M}_n(o) = 10^5$ and $c = 10$ g·dm^{-3} after 10 main chain scissions the concentration of cyclobunanol structural units is about 10^{-4} mol·dm^{-3} in the main chain. Such a low concentration is difficult to prove. Their presence has not been proved after extensive irradiation in homopolymers and copolymers either. Moreover, the formation of a cyclobutane ring as a part of the main chain may be more difficult in polymers than in low molecular mass compounds.

Since the photochemical reaction originates from the triplet state, another source of possible error is the presence of oxygen which is a known triplet quencher. There are several more or less efficient methods for removal of oxygen from solutions. Usually freeze-thaw cycle is preferred but the majority of quantum yields in solution were obtained under the conditions of bubbling nitrogen through the solution. It is even more difficult to keep an inert atmosphere during irradiation in film. Table 3 contains the values of quantum yields in de-aerated and aerated solutions. These data indicate that it is necessary to de-aerate the solution for derivatives with the long lifetime of the triplet state. This conclusion was confirmed by the irradiation in film as well [35].

The quantum yields of main chain scission of poly(1-/4-X-phenyl/-2-propen-1-one) in a solution of a benzene are summarized in Table 3. Most photolysis were performed at 366 nm, which is the long wavelength edge of the n-π^* band. The data obtained at 313 nm are the same as those at 366 nm. No wavelength dependence was observed. Since several sources of errors are involved, the quantum yields are loaded with error ($\pm 50\%$ or even higher).

The quantum yields of main chain scission are about 0.3 for reactive polyketones and up to 10^{-4} for the less reactive ones. As Scheme 1 indicates, the Norrish type II reaction is a two-step process involving γ-hydrogen abstraction with the triplet and β-bond carbon-carbon scission in the triplet 1,4-biradical. Photoreactive ketones having a triplet state with the n-π^* character, form the 1,4-biradical with unit efficiency. 1,4-Biradical collapses with the formation of the products and of the original compound in the ground state. The quantum yield is not therefore directly related to triplet reactivity [31].

Polyketones with a resolved n-π^* band in UV spectrum and well resolved emission at 77°K exhibit quantum yields in the region 0.2–0.4. On the other hand, the quantum yields are low (about 10^{-4}) for polyketones with no distinct n-π^* UV band and broad unresolved emission at 77°K. Transition between reactive and unreactive polyketones is represented by poly(1-/4-methoxy phenyl/-2-propen-1-one) with a quantum yield around 0.1 which is strongly dependent upon the polarity of solvent [18]. The influence of solvent polarity on the course of reaction is more difficult to study with polymer substrates because of the problems with polymer solubility and solution characterization of polyketone–solvent pair.

The quantum yields of main chain scission for copoly-

Table 3. Photolysis of poly(1-(4-X-phenyl)-2-propen-1-ones) in solution [6].

Polymers	$[\eta] \times 10^{2}$ [a] ml·g^{-1}	$\overline{M}_n \times 10^{5}$ [c]	$\overline{M}_w/\overline{M}_n$ [f]	Ø [k] N$_2$	Ø [k] Air	K_{SV} [p] dm^3·mole^{-1}
Poly(1-phenyl-2-propen-1-one)	0.55	1.10[d]		0.3		
	0.82	2.19[e]		0.22	0.22	68
	0.78	2.94[f]		0.19[e]		64
		0.67[f]		0.245[m]		
		2.80[f]	1.65	0.4–0.6[n]		90
	0.71	1.95	2.6	0.24		82
Poly(1-(4-methylphenyl)-2-propen-1-one)	1.04	1.19[f]	3.72	0.26	0.21	436
Poly(1-(4-ethylphenyl)-2-propen-1-one)	3.20	10.1[g]		0.25	0.22	345
Poly(1-(4-tercbutylphenyl)-2-propen-1-one)	1.98	3.02[fr]	2.90	0.37	0.32	336
Poly(1-(4-cyclopentylphenyl)-2-propen-1-one)	0.74	1.11[f]	2.42	0.33	0.30	386
Poly(1-(4-fluorophenyl)-2-propen-1-one)	0.94	2.42[h]			0.31	62
Poly(1-(4-chlorophenyl)-2-propen-1-one)	1.34	4.5[g]			0.39	260
Poly(1-(4-bromophenyl)-2-propen-1-one)	0.32	0.67[h]			0.36	110
Poly(1-(4-iodophenyl)-2-propen-1-one)[o]	0.15[b]					
Poly(1-(4-methoxyphenyl)-2-propen-1-one)	2.21	6.50[i]		0.10[l]		1700[l]
Poly(1-(3,4-dimethoxyphenyl)-2-propen-1-one)	0.71	1.74[h]		1.1×10^{-4}		
Poly(1-(4-biphenyl)-2-propen-1-one)		1.14[i]		2.6×10^{-4}		
Poly(1-(2-naphthyl)-2-propen-1-one)		0.11[i]		—		
Poly(1-(4-acetylphenyl)-2-propen-1-one)	0.54[b]	2.0[j]	3.2		0.22[r]	381[r]

[a]Limiting viscosity number, benzene, 30°C.
[b]Dioxane, 30°C.
[c]Number average molecular weight.
[d]Viscosimetrically, benzene, 30°C, $\alpha = 0.7$, $K = 10^{-2}$, $\overline{M}_n = M_v/2$.
[e]Osmometrically, benzene, 30°C.
[f]Calculated from GPC curve, THF, using polystyrene as standard.
[g]Osmometrically, toluene, 37°C.
[h]Viscosimetrically, benzene, 30°C, $\alpha = 0.84$, $K = 2.82 \times 10^{-3}$.
[i]Osmometrically, chlorobenzene, 30°C.
[j]Osmometrically, dioxane, 30°C.
[k]Quantum yield of main chain scissions at 366 nm radiation in benzene.
[l]313 nm, chlorobenzene.
[m]313 nm, benzene.
[n]347 nm, irradiation with pulses.
[o]No main chain scission but elimination of J_2.
[p]Quenching of photolysis with naphthalene in benzene.
[r]Dioxane.

Table 4. Photolysis of copolymer 1-(4-carbethoxy phenyl)-2-propen-1-one (PCOOEt) with styrene (S), phenyl acrylate (PhA), ethyl acrylate (EA) and acrylonitril (AN) in benzene solution at 366 nm [16].

Polymer	$I_o \times 10^4$ E·dm^{-3}·min^{-1}	$\epsilon \cdot c \cdot d$ mol·dm^{-3}·cm^{-1}	$\overline{M}_n$ [a] $\times 10^{-3}$ Dalton	c g·dm^{-3}	S mol·min^{-1}	Ø mol·E^{-1}	K_{SV} mol^{-1}	k_r [b] $\times 10^{-7}$ s^{-1}
PCOOEt	0.459	0.800	650	4.00	2.150	0.45	170	1.18
PCOOEt/S-A	0.988	0.061	259	3.77	0.260	0.29	172	1.16
PCOOEt/S-B	0.988	0.112	242	3.98	0.303	0.22	167	1.20
PCOOEt/S-C	0.988	0.222	215	3.97	0.825	0.39	173	1.16
PCOOEt/S-D	0.988	0.303	197	3.62	0.985	0.34	175	1.14
PCOOEt/PhA	0.997	0.029	97	10.08	0.0044	0.07	280	0.71
PCOOEt/EA	0.988	0.036	580	3.09	0.670	0.45	480	0.42
PCOOEt/AN[c]	1.196	0.120	153[d]	4.11	0.0255	0.024	170	—

[a]Number average molecular weight determined by GPC.
[b]Calculated for $k_q = 2 \times 10^{9}$ mol^{-1}·s^{-1}.
[c]Dimethylformamide.
[d]Mark-Houwink coefficients $K = 16 \times 10^{-3}$ ml·g^{-1}, $\alpha = 0.81$ at 25°C.

Table 5. Initial quantum yield of main chain scission of poly(1-(4-X-phenyl)-2-propen-1-one) photolyzed in film by 313 nm [35].

X	$c \times 10^4$ g/cm²	S_i^a s⁻¹	$I_A \times 10^{10}$ E·cm⁻²·s⁻¹	$\overline{M}_w^b(o) \times 10^{-5}$	$\overline{M}_w(o)/\overline{M}_n(o)^c$	$\varnothing_i^d$	T_g^e °C
H	3.438	0.04	7.87	5.21	4.17	0.14	73.5
4-F	5.093	0.04	9.17	6.90	4.72	0.15	
	1.273	0.07	11.0	6.90	4.72	0.06[d]	
4-Cl	3.183	0.05	11.84	10.12	3.75	0.05	89
4-C_2H_5	3.119	0.05	7.58	13.28	3.68	0.06	51.5
4-OCH_3	3.271	0.004	24.76	19.20	3.99	0.001	

[a]Rate of degradation; number of main chain scission during time.
[b]Determined by LALLS.
[c]Determined by GPC.
[d]$\varnothing_i = c \cdot S_i/\overline{M}_n(o)$, $\overline{M}_n(o)$ was calculated using $\overline{M}_w(o)$ determined by LALLS and $\overline{M}_w(o)/\overline{M}_n(o)$ by GPC.
[e]Reference [41].

mers are nearly the same as for homopolymers [16]. Additional sources of error for quantum yields are chemical inhomogeneity and polydispersity [16]. A comparison of quantum yields of main chain scission in copolymers leads to the conclusion that they are lower when the substituents in γ-position are electron donors (Table 4). The same conclusion is valid for the low molecular mass compounds [31].

The plots of the number of main chain scissions against time are clearly curved in the case of the photolysis in film [35]. This curvature is not observed in the photolysis of homopolymers and copolymers in solution. Quantum yields of main chain scission in film calculated from the initial part are lower than those in solution (Table 5). Polyketones investigated elsewhere [35] have T_g higher than the laboratory temperature [41] (Table 5). Consequently, the mobility of 1,4-biradical in glassy matrix is lower than in solution. The formation of the original structure is preferred under these conditions and the quantum yield of main chain scission is lower than in solution [42].

QUENCHING OF PHOTOLYSIS

The typical triplet quenchers, with the triplet level lower than that of aryl alkyl ketones, quench phosphorescence and transients and inhibit photolysis in solution and in film. The extent of degradation decreases in the presence of triplet quenchers. Using viscometric monitoring the ratio of quantum yields without and with a quencher is equal to the number of main chain scission without and with a quencher. Since this measurement is relative, the error involved is lower than that for quantum yields. The Stern-Volmer model quantitatively describes the inhibition of photolysis in solution resulting from the physical transfer of the triplet energy. The Stern-Volmer model states that:

$$\phi_o/\phi = 1 + K_{SV}[Q]$$

where K_{SV} is the Stern-Volmer constant and $[Q]$ is the concentration of the quencher [2,43]. The Stern-Volmer constant is given by the product of the lifetime of the excited state in the absence of quencher (τ) and the bimolecular quenching rate constant (k_q). If one of these parameters is known the second may be calculated. Static measurements of this type can yield the data which are otherwise accessible by the sophisticated pulse techniques [2,25,26,28,29]. Both the lifetime and the bimolecular rate constant for quenching are influenced by several factors. Several values of K_{SV} for systems of polymer-low molecular mass quencher (naphthalene, biphenyl) are given in Table 2. As is evident from Figure 6, K_{SV} more or less reflects the changes in the lifetime although k_q changes as well.

The bimolecular rate constant for quenching is often approximated by the diffusion-controlled rate constant [2,43]. For benzene (or chloroform) at 25°C k_{diff} equals

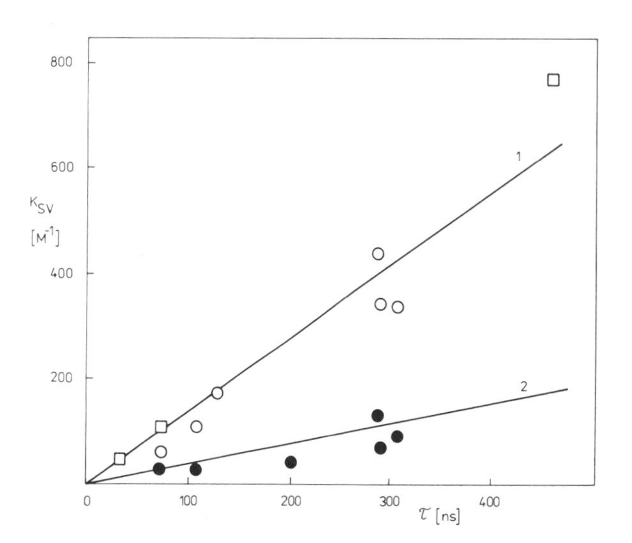

Figure 6. Plots of K_{SV} vs. τ taken from Table 2. K_{SV} for the quenching with naphthalene for homopolymers (1, open circles) and copolymers with styrene (1, open squares) and for homopolymers quenched by biphenyl (2, full circles).

1.10^{10} $dm^{-3} \cdot mol^{-1} \cdot s^{-1}$. The measurements of model compounds and polymers indicate that the bimolecular quenching rate constant is always lower than k_{diff} [25,44, 45]. Comparison of k_q for model compounds, homopolymers, and copolymers follows the order: k_q/model $>$ k_q/copolymer $\geq$ k_q/homopolymer (Table 2). The greatest difference in intermolecular quenching of the model and the polymer is observed for 1-/3,4-dimethoxyphenyl/-2-propen-1-one chromophore which has the longest triplet lifetime. The quenching rate constants for copolymers are usually a factor of 3 less than for model compounds. This reduced efficiency is often attributed to the low rate of self-diffusion of the polymer bound chromophore. However, even if the chromophore were assumed to be stationary on the time scale of the experiment, this reduced efficiency would introduce a factor 2. Thus we come to the conclusion that the presence of polymer chain restricts access to the chromophore, so that a greater proportion of collisions between the small molecular quencher and the excited chromophore are ineffective. The ratio (k_q/model)/(k_q/homopolymer) is either the same or even higher than that of copolymer. If there were an effective migration along the polymer chain, then this ratio should be at least lower than that of copolymer or even lower than 1. This means that the quenching of homopolymers should be more efficient. Inevitably, one of these conclusions occur:

1. There is no migration of the triplet energy along the polymer chain.
2. The efficiency of intermolecular quenching is decreased due to rapid intramacromolecular energy transfer to energetic traps.

Both K_{SV} and k_q indicate that the triplet quenchers naphthalene, biphenyl, conjugated dienes, and 4-acetyloxy-2,2,6,6-tetramethylpiperidine-N-oxyl exhibit different quenching ability.

Triplet quenching in a macromolecule can take place as an inter- or intramacromolecular process depending on whether the quencher is free or covalently bound to the polymer. We have shown earlier that the bound quenchers are more efficient in solution than the free ones [32,46,47]. The reason for this effect is in the higher local concentration of the quencher in the polymer coil that can be achieved for bound quenchers. Photolysis and quenching studies of poly(1-/4-methoxyphenyl/-2-propen-1-one) indicated the involvement of the long-lived triplet [18]. Copolymers of (1-/4-methoxyphenyl/-2-propen-1-one) with functionalized triplet quenchers were prepared and their photolysis was studied in solution [46,47]. The results from these studies were not conclusive because it was not possible to extract from these data, based on viscometry, the values of (k_q/free quencher) and (k_q/bound quencher) for comparison. The transient spectra may be more informative in this respect. Transient spectra of copolymers with a bound

Table 6. Decay of transient absorptions in 1-(4-methoxyphenyl)-2-propen-1-one copolymers with functionalized triplet quenchers in chloroform at room temperature [25].

Quencher	Content wt./wt.%	$[\eta]^d$ ml·g^{-1}	$C_L^e \times 10^3$ (M)	λ_{max} nm	$k \times 10^6$ s^{-1}
2-NMA[a]	4.20	125	2.85	600	4.79
	2.25	209	0.91	610	1.94
	2.40	150	1.36	650	2.27
BMA[b]	2.5	103	1.83	600	3.07
	3.3	208	1.20	600	2.27
	4.7	147	2.40	620	5.34
	1.8	169	0.80	620	1.77
	4.9	201	1.84	620	5.34
	2.95	199	1.11	620	4.68
BPMA[c]	1.05	240	0.24	600 390	0.72 0.75
	3.54	229	0.85	390 600	1.20 1.62
	4.57	236	1.07	390	1.56
	6.60	247	1.47	600	2.20
	8.25	265	1.72	600 390	3.60 3.60

[a]Naphthyl methacrylate (2-NMA).
[b]Biphenyl methacrylate (BMA).
[c]2-Hydroxy(4-(2-methacroyloxyethoxy)benzophenone) (BPMA).
[d]Ethyl benzoate.
[e]Local concentration of quencher at total concentration of copolymer 1.5 g/l.

quencher are more complex than those of the respective homopolymers. In addition to absorptions at 390 and 620 nm belonging to (1-/4-methoxyphenyl/-2-propen-1-one) they show absorptions at 420 and 350 nm for (2-naphthyl methacrylate /2-NMA) and at 350 nm for (p-biphenyl methacrylate /BMA). The growth of the transient absorption at 420 nm for 2-NMA was very rapid. The rate constant for this growth for BMA was $k = 2.5 \times 10^1$ s^{-1} and did not depend on bulk concentration. Both maxima (390 and 620 nm) were used to monitor decay of the transient of the copolymer with (2-hydroxy-/4-/2-methacroyloxy/ethoxy/benzophenone /BPMA). The local concentration was calculated according to [6]:

$$c_L = (1.2 \times 10^3)m \cdot w/[\eta]M$$

where w is the weight fraction of the quencher in copolymer, m is the copolymer concentration in solution in $g \cdot dm^{-3}$, $[\eta]$ is the limiting viscosity number and M is the molecular mass of the structural unit.

The data summarized in Table 6 indicate that the dependence of the decay rate constant on the bound quencher is linear (Figure 7). Since the rate constant for decay is low (3.4×10^4 s^{-1}) the abscissa of the plot k vs. c_L is rather small. Though the scatter of the experimental points is rather high, the slope yields an apparent value for $k_q = 1.25 \times 10^9$ $dm^{-3} \cdot mol^{-1} \cdot s^{-1}$. This value agrees well with k_q for the intermacromolecular quenching of the copolymer (1-/4-methoxy-phenyl/-2-propen-1-one)

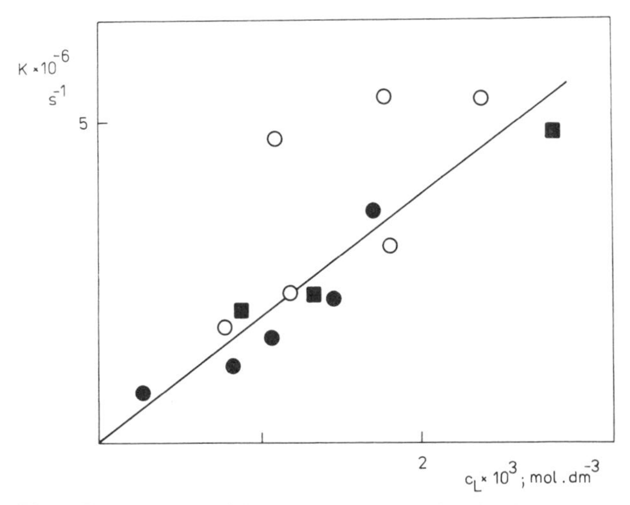

Figure 7. Dependence of the decay rate constants *k* on local concentration of quenchers for copolymers 1-(4-methoxy-phenyl)-2-propen-1-one with 2-NMA (full squares), BMA (open circles), and BPMA (full circles).

and methyl methacrylate with diene (Table 2). All bound quenchers show similar efficiency. Through the approximate analysis indicated above it is possible to conclude that bound quenchers are not intrinsically more efficient that the free ones. The same will apply if a chemical reaction of the quencher is to occur instead of the dissipation of the energy. This means that inter- and intramacromolecular triplet sensitization is by no means more efficient than with a small molecule. Moreover, we cannot expect substantially higher reactivity in a triplet reaction centre of this type.

CONCLUSIONS

Polymers with aryl alkyl carbonyls as side group chromophores are not commercially available at present with the exception of some copolymers with a low carbonyl content used as packaging materials. In the future their importance will probably grow mainly because of their application as photoresists for microelectronic, photoinitiators, photosensitizers and photocross-linking agents. In this contribution we have pointed out some aspects of the Norrish type II reaction, especially the differences due to macromolecular structure. Specific features are very pronounced when the carbonyl monomers are bound with other monomers or quenchers in a macromolecule. The processes which occur in macromolecules as a result of binding different chromophores are qualitatively fully understood. For a more detailed study, homopolymers and copolymers with better defined structures are needed.

The fate of the excitation of carbonyl groups in polymers after the first 10 ns is clear. There are, however, some problems with the resolution of the short-lived triplet from 1,4-biradical intermediates at self-quenching and energy transfer. The picosecond spectroscopy is to be employed to elucidate the fate of excitation after absorption until the formation of 1,4-biradical. This information will help to settle the question of isothermal energy transfer along the polymer chain.

The photochemical data reviewed in this article allow us to make the following conclusions concerning the photochemical behaviour of isolated carbonyl chromophore in carboneous polymers:

1. The Norrish type II reaction will occur for short-living triplets in film with a quantum yield lower than that in solution.
2. For the long-lived triplets and accessibility of oxygen, the reaction will be quenched forming singlet oxygen which exhibits its own photochemistry.
3. Quenching of the triplet state of isolated chromophores by inter- and intramolecular processes at the concentrations typical of light stabilizers is not an efficient process for long-term stabilization.

REFERENCES

1. Calvert, J. G. and J. N. Pitts, Jr. *Photochemistry*. New York:J. Wiley and Sons, Inc., p. 377 (1966).
2. Turro, N. J. *Modern Molecular Photochemistry*. Benjamin Cummings Publishing Company, Inc., p. 363 (1978).
3. Ranby, B. and J. F. Rabek. *Photodegradation, Phtooxidation and Photostabilization of Polymers*. New York:J. Wiley and Sons (1975).
4. Guillet, J. E. and R. G. W. Norrish. *Proc. Roy. Soc.*, London, 233A:153 (1955).
5. Wissbrun, K. F. *J. Am. Chem. Soc.*, 81:58 (1958).
6. Hrdlovič, P. and I. Lukáč. *Developments in Polymer Degradation*, 4:101 (1982).
7. Guillet, J. E. *Polymer Photophysics and Photochemistry*. Cambridge, UK:Cambridge University Press, Chapter 10 (1985).
8. Daly, W. H. *Encyclopedia Polymer Science and Technology*, 14:617 (1971).
9. Merle-Aubry, L., Y. Merle and E. Selegny. *Makromolekulare Chemie*, 176:709 (1975).
10. Hrdlovič, P., J. Trekoval and I. Lukáč. *European Polymer J.*, 15:229 (1979).
11. Yokota, K., T. Suzuki, S. Nakazawa, K. Nakamura and Y. Takada. *Hokkaido Daigahn Kogabubu Kenkyn Hokoku*, 60:63 (1971); *Chem. Abs.*, 75:110888b (1971).
12. Otsu, T. and H. Tanaka. *Polymer*, 16:468 (1975).
13. Sastre, R., J. L. Acosta, R. Garrido and J. Fontan. *Makromol. Chemie*, 63:85 (1977).
14. Nenkov, G., T. Bogdantsaliev, T. Georgieva and V. Kabaivanov. *Polymer Photochem.*, 6:475 (1985).
15. Aglieto, M., L. Cioni, G. Ruggeri, E. Taburoni and E. Tomei. *J. Polymer Sci., Part A, Polymer Chem. Ed.*, 24:1337 (1986).
16. Lukáč, I., E. Mikulová and P. Hrdlovič. *European Polymer J.*, 23:929 (1987).

17. Lukáč, I., Š. Chmela and P. Hrdlovič. *Polymer Sci., Polymer Chem. Ed.*, 17:2893 (1979).
18. Lukáč, I., M. Moravčík and P. Hrdlovič. *J. Polymer Sci., Polymer Chem. Ed.*, 12:1913 (1974).
19. Lukáč, I., J. Pilka, M. Kulícková and P. Hrdlovič. *J. Polymer Science, Polymer Chem. Ed.*, 15:1645 (1977).
20. Lukáč, I. and P. Hrdlovič. *European Polymer J.*, 14:339 (1978).
21. Lukáč, I., S. Chmela and P. Hrdlovič. *J. Photochemistry*, 11:301 (1979).
22. Hrdlovič, P., J. Daněček, D. Berek and I. Lukáč. *European Polymer J.*, 13:123 (1977).
23. Lukáč, I. and P. Hrdlovič. *Polymer Photochemistry*, 2:277 (1982).
24. Lukáč, I., P. Hrdlovič and J. E. Guillet. *Polymer Photochemistry*, 7:163 (1986).
25. Hrdlovič, P., J. C. Scaiano, I. Lukáč and J. E. Guillet. *Macromolecules*, 19:1637 (1986).
26. Hrdlovič, P., G. Guyot, J. Lemaire and I. Lukáč. *Polymer Photochemistry*, 3:119 (1983).
27. Hrdlovič, P. *Polymer Photochemistry*, 7:359 (1986).
28. Faure, J., J. P. Fouassier, D. J. Loungnot and P. J. Salvin. *J. Nouv. Chem.*, 1:15 (1977).
29. Kiwi, J. and W. Schnabel. *Macromolecules*, 9:468 (1976).
30. Chapman, O. L. and G. Wampfler. *J. Am. Chem. Soc.*, 91: 5390 (1969).
31. Wagner, P. J. *Accounts Chem. Res.*, 4:168 (1971).
32. Lukáč, I., P. Hrdlovič, Z. Maňásek and D. Belluš. *J. Polymer Sci.*, A-1(9):69 (1971).
33. Beck, G., J. Kiwi, D. Lindenau and W. Schnabel. *European Polymer J.*, 10:1069 (1974).
34. Golemba, F. J. and J. E. Guillet. *Macromolecules*, 5:212 (1972).
35. Hrdlovič, P. and J. E. Guillet. *Polymer Photochemistry*, 7: 423 (1986).
36. Lewis, F. D. and T. A. Hilliard. *J. Am. Chem. Soc.*, 84: 3852 (1972).
37. Wagner, P. J. and A. E. Kemppainen. *J. Am. Chem. Soc.*, 94:7495 (1972).
38. Salvin, R., J. Meybeck and J. Faure. *J. Photochemistry*, 6:9 (1976,1977).
39. Casals, P. J., J. Ferard, R. Rupert and M. Keravec. *Tetrahedron Letters*, 45:3909 (1975).
40. Bays, J. P., M. V. Encinas, R. D. Small, Jr. and J. C. Scaiano. *J. Am. Chem. Soc.*, 102:727 (1980).
41. Pizzarini, G., P. Magagnini and P. Giusti. *J. Polymer Sci.*, 9(A-2):1133 (1971).
42. Dan, E. and J. E. Guillet. *Macromolecules*, 6:230 (1973).
43. Turro, N. J. *Pure Appl. Chem.*, 49:405 (1977).
44. Beck, G., G. Dobrowolski, J. Kiwi and W. Schnabel. *Macromolecules*, 8:9 (1975).
45. Scaiano, J. C., E. A. Lissi and L. C. Stewart. *J. Am. Chem. Soc.*, 106:1539 (1984).
46. Lukáč, I. and P. Hrdlovič. *European Polymer J.*, 11:767 (1975).
47. Lukáč, I. and P. Hrdlovič. *European Polymer J.*, 15:533 (1979).

B. J. Tighe[1]

Applications of Controlled Degradation in Biomaterials Research

ABSTRACT

Two aspects of degradation processes that are of interest in current biomaterials research are discussed. In both cases an underlying appreciation of structure and mechanism are necessary in order to achieve the degree of control in the degradation processes that will enable more effective devices to be designed. The first area is the hydrolytic degradation of ester-containing polymers, in particular hydroxybutyrate-hydroxyvalerate copolymers. The importance of surface-mediated events, particularly in the early stages of the degradation (often referred to as bioerosion) is discussed. Particular emphasis is placed upon the role of goniophotometry as an experimental technique in this type of study. The goniophotometer (a non-destructive method of optical monitoring) is also valuable in the second area discussed, the use of oxygen plasmas in biomaterials research. Principal applications discussed here involve harnessing the unusual behaviour of aromatic rings in device fabrication and in the preparation of tissue culture substrates.

KEY WORDS

Biomaterials, bioerosion, biodegradation, goniophotometry, oxygen plasma, polymer surfaces.

INTRODUCTION

There are several ways in which the control of degradation processes governs the success of biomaterials research. Examples of two distinct aspects will be discussed here. The first category centres around the design of polymers to undergo degradation at a controlled rate in a biological environment. Perhaps the best known illustration of this technique is the use of polymers as biodegradable, or bioabsorbable, sutures. The second category involves the use of controlled degradation as part of a fabrication process, perhaps in modifying the surface of a polymer for more effective performance at a biological interface.

In discussing bioerosion, it is appropriate to describe a technique that has been of great use in our work. It is of considerable relevance at a conference such as this one because of its applicability to a range of degradation studies, although the lack of commercial instrumentation obviously restricts its use. It is particularly sensitive in the study of the vital early stages of degradation processes in which surface events are important. The technique, which involves the study of light scattered from the sample surface, is known as goniophotometry.

THE USE OF GONIOPHOTOMETRY FOR BIOEROSION AND RELATED STUDIES

Background to the Technique

The principle of goniophotometry was first used in our work in a relatively unsophisticated manual device based on a Brice-Phoenix light scattering photometer. Readings were taken from a galvanometer and a single sample took some two hours to scan. Plotting the results together with calculation of curve parameters took a further hour or so. An intermediate motor driven device, using the same light box and linked to a chart recorder, was followed by an instrument in which a more sophisticated motor drive is linked (via a Peterson a/d converter) to a Sharp MZ80K microcomputer. This instrument, designed and constructed at Aston, can scan a sample and plot a curve with calculations in some ten minutes.

Our current work involves modification of the existing instrument in several important respects. These are:

1. Provision of five monochromatic illumination wavelengths of sufficient intensity for use with low reflectance samples. A mixture of filtered lamps and monochromatic spectral line lamps are envisaged. The purpose of wavelength variation is to enable investigation of the phenomena associated with peak broadening and asymmetry which are able to provide a more complete characterisation of the important early stages of the degradation processes.

[1]Speciality Polymer Group, Aston University, Birmingham B4 7ET UK.

2. Development of a series of sample mounts, based on fibre-optic manipulators, that will enable samples to be adjusted remotely whilst in the light beam. A particular example that illustrates the difficulties involved is a polymer fibre. Although fibres and rods have characteristic scattering envelopes they are quite suitable for study.
3. Alternative microcomputers with greatly improved graphics and facility for multiple screen overlays, mouse control for curve rescaling/manipulation and with improved data handling speed have greatly enhanced the technique and enable full use of wavelength variation to be made. The BBC Master model 512 and the Cifer have both proved to be useful in this respect.

Initial work was carried out on pigmented paint films and specifically the early stages of their deterioration or ageing of surface coatings. Subsequently use of the instrumentation has begun in the study of a variety of surface modification processes involving polymers. These include the reaction of polymers with gas plasmas, polymer modification at biological interfaces and, most recently, the hydrolytic degradation of potentially bioerodible polymers (notably a series of recently introduced hydroxybutyrate-hydroxyvalerate copolymers). It is in areas such as these that there is a need for nondestructive techniques which give information about the important early stages of polymer modification processes. Goniophotometry is just such a technique but largely because of the lack of available instrumentation its potential is virtually unexploited.

GONIOPHOTOMETRY: PRINCIPLES AND INSTRUMENTATION

The background to this subject is found in physics and is concerned with the scattering of light from solid surfaces. This scattering manifests itself in what is known as gloss—which is the reason for the connection of this whole subject with surface coatings. Far from being a trivial phenomenon, the gloss of a surface is both difficult to define and measure with any degree of precision. Despite the complications encountered in defining the way in which a human subject assesses a surface, there has been considerable progress in characterisation of reflectance by instrumental techniques and in relating the scattering to fundamental physical phenomena. It is in this context that the goniophotometer was developed. It could be described as a technique based on the physical phenomena and designed as a means of quantifying subjective surface assessment. The potential of the technique goes far beyond this, however, and it is with the further exploitation of the technique that our present work is concerned. The historical and current positions relating to goniophotometry and the associated physical phenomena are described in a review contained in *Polymer Surfaces* (D. T. Clark and W. J. Feast, editors) [1]. The outline that follows is based on that review. The relevant terms associated with the field are illustrated in Figure 1. This shows how an incident light beam, is reflected from a mirror as a specular beam [Figure 1(a)], how the scattering envelope appears for idealised diffuse reflectance [Figure 1(b)] and for a more typical surface from which both diffuse and specular reflectance occur [Figure 1(c)]. In principle these scattering envelopes may be determined with an experimental technique containing the elements shown diagrammatically in Figure 2.

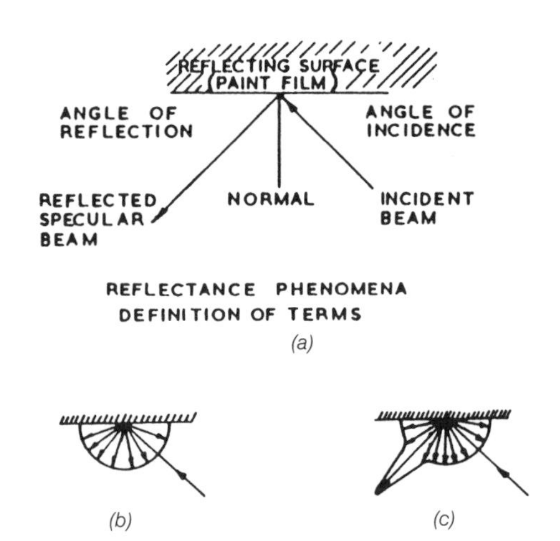

Figure 1. Reflectance phenomena: (a) definition of terms; (b) spatial distribution of light reflected from an ideal matt surface; (c) spatial distribution of light reflected from a semi-matt surface.

It is significant that the attempts to harness the principle (which are reviewed in Reference [1]) have only included one commercial instrument and have been exclusively (apart from work in these laboratories) concerned with paints. The work that we have done has been initially based on the Brice-Phoenix light scattering photometer and benefits from the fact that this is designed to measure the intensity of light scattered from polymer so-

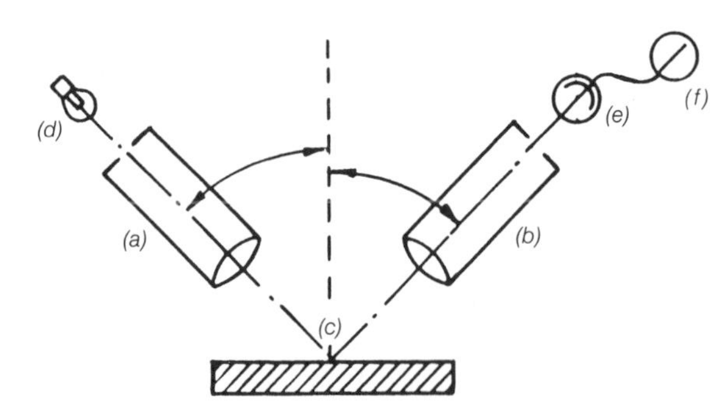

Figure 2. Essential features of goniophotometer. Components: (a) collimating lens system for incident beam from light source (d); (b) collimating lens system for reflected beam; (e) and (f) represent the detector and read-out meter respectively. The incident and reflectance angles may be varied about point (c).

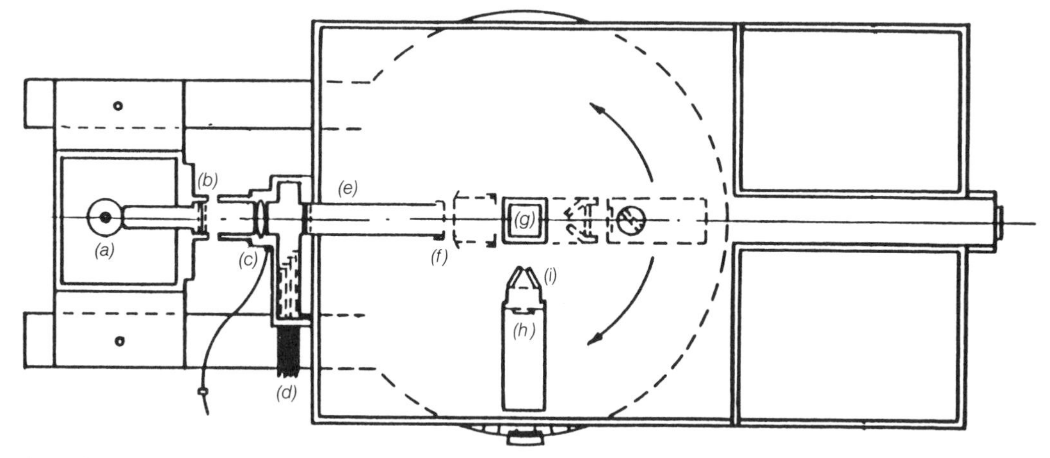

Figure 3. Essential features of a high-performance goniophotometer, based on a Brice-Phoenix light-scattering photometer.

lutions in comparison to the intensity of the incident beam. The wide range of intensities involved is handled by a series of neutral density filters that avoid the need to rely on the accuracy of electronic scaling over the intensity range. The mechanics and electronics are relatively straightforward and fabrication from scratch does not present any problem for the well-equipped workshop. Figure 3 shows the layout of the goniophotometer design which we have used. Light from lamp (a) passes through an optional wavelength filter (b) and shutter unit (c) into a collimating tube (e) with variable slit aperture (f). The light impinges on the sample mounted at (g) and the scattering envelope is determined by measuring intensity as a function of receiving angle at the rotating photocell (h) which has a variable slit aperture (i). The intensity of the incident beam is controlled by neutral density filters (d) and the intensity of the reflected beam is amplified from the photocell and recorded or processed. Typical goniophotometric curves obtained with the above instrumentation are shown in Figures 4 and 5. These correspond to an undegraded glossy black pigmented film (Figure 4) and a dramatically degraded matt version (Figure 5). Each figure shows the scattering envelope for four different angles of incidence (30–75′).

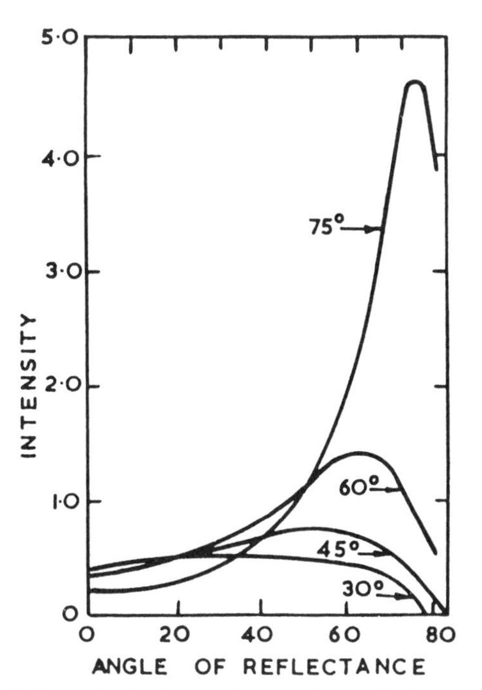

Figure 4. Typical goniophotometric curve for black pigmented polymer film.

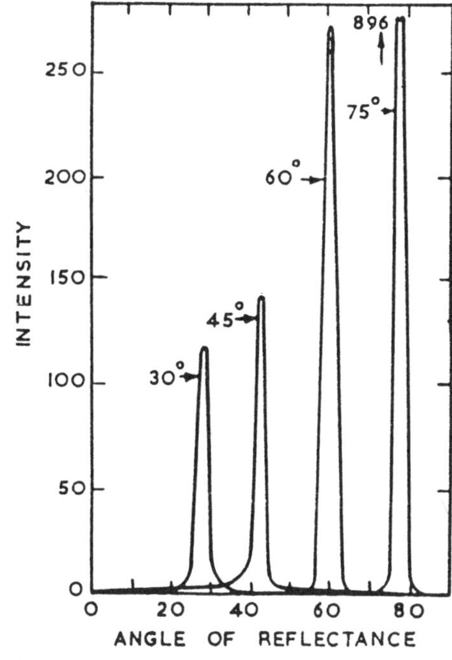

Figure 5. Typical goniophotometric curve for black pigmented polymer film.

The sensitivity of the technique is illustrated by the fact that the initial intensity (1000 units) is the same in both cases and yet good full scale scattering envelopes with maximum intensities of less than ten units are obtained (Figure 5). This is because of the neutral density filter range, and in fact, a further five-fold increase in sensitivity beyond that shown in Figure 5 is available without further electronic signal amplification. For this reason surface scattering from unpigmented films is well within the scope of the instrument and enables any suitably presented polymer sample to be scanned. This aspect of the use of the technique has been little explored as yet, however.

INTERPRETATION OF GONIOPHOTOMETRIC CURVES: BIOEROSION STUDIES

The ways in which the scattering envelope may be characterised are described in Reference [1]. Briefly, the nature and dimensions of surface rugosity affect the pattern of scattered light and thus the shape of goniophotometric curves. Micro defects which are not greater than the wavelength of the incident light beam remove energy from the beam—principally by Mie-type scattering. In this region peak height and area diminish, but the shape of the peaks remain substantially unchanged. As the size of imperfections increases, the scattering envelope produces asymmetric peaks—a feature illustrated in extreme form by Figure 5. This relationship of surface rugosity to peak asymmetry is an important feature of the technique.

The scattering envelope is most conveniently characterised by three parameters. These parameters are the intensity of light reflected at the specular angle (Is), the diffuse reflectance measured normal to the specimen (Id) and the peak width at half height ($W_{1/2}$). These three factors are combined together in the so-called gloss factor, given by $(Is - Id)/W_{1/2}$. These parameters are illustrated in Figures 6 and 7 in relation to the hydrolytic degradation of poly(hydroxybutyrate-hydroxyvalerate) copolymers, sold under the tradename *Biopol* by Marlborough Biopolymers/I.C.I. Both are injection moulded samples of the 20% hydroxyvalerate copolymer with an initial molecular weight (*Mw*) of 300,000. Whereas that shown in Figure 6 is undegraded, however, Figure 7 shows the curve for a sample that has been maintained in aqueous buffer at pH 7.4 and 37°C for 5500 hours. The fairly symmetrical peak of Figure 6 corresponds to a relatively smooth surface, using the criteria outlined above. Development of surface imperfections smaller than the wavelength of the light beam used (=0.5 μm) lead to a removal of energy from the beam and a reduction in peak intensity with little change in peak asymmetry. It is clear, however, that something more has occurred in Figure 7. Here the peak width at half-height has increased together with the magnitude of *Id*. This change corresponds to an increase in surface rugosity and the development of imperfections whose dimensions are greater than 0.5 μm. Work on the correlation (for paint surfaces) of surface features using interferometry, surface profilometry and scanning electron microscopy is described in Reference [2] and references therein.

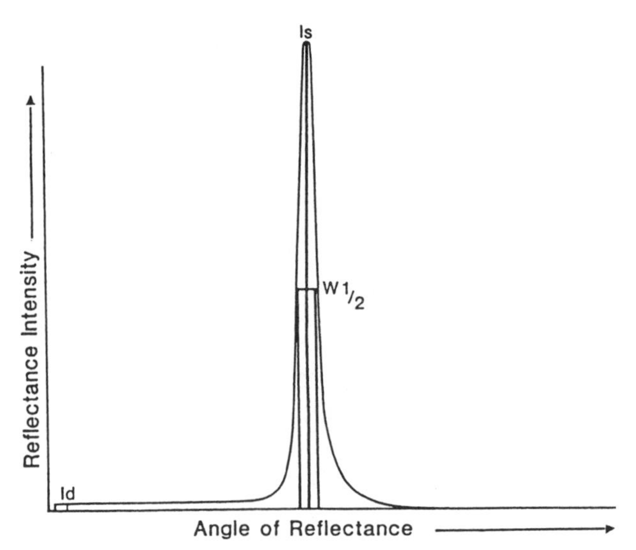

Figure 6. Goniophotometric curve of an injection moulded sample of hydroxybutyrate-hydroxyvalerate copolymer (12% PHV, *Mw* 30,000).

The important feature associated with Figures 6 and 7 is that there is currently considerable debate as to the extent to which these polymers degrade in physiological conditions. This was reflected in papers at a recent biomaterials conference [2]. In fact weight loss measurements show no statistically significant erosion of the sample corresponding to Figure 6 whereas goniophotometry indicates that a marked change has taken place.

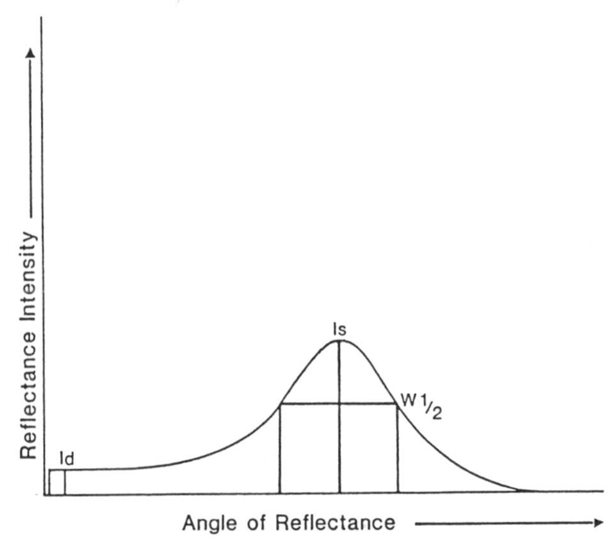

Figure 7. Goniophotometric curve of sample corresponding to Figure 6 after immersion in an aqueous buffer at 37°C and pH 7.4 for 5500 hrs.

Our studies lead us to believe that surface changes are an important precursor and indicator of the more dramatic degradation that eventually occurs with these materials.

THE BIOEROSION PROCESS: AN OVERVIEW

The range of applications in which bioerosion is important includes surgical fixation and drug delivery devices. A recent overview of literature on the available materials for such work summarises the ester-containing polymers in the form shown in Table 1 [3].

A second paper in the same series summarises the contributing effects that enable some control over the rate of bioerosion or biodegradation to be exercised [4]. They include molecular structure, crystallinity, molecular weight and fabrication process (although these may be interdependent to some extent).

Detailed studies of hydroxybutyrate-hydroxyvalerate copolymers, in comparison with other polyesters, provide some insight into the general mechanism of bioerosion. The relative effects of pH variation in comparison with those of differing biological fluids indicate that the underlying mechanism is usually hydrolysis, rather than true biodegradation. Having said that, specific instances of bacterial and enzymatic degradation do occur.

Gravimetric measurements taken together with surface energy, molecular weight and goniophotometry measurements provide an interesting composite picture of the degradation process. In the early stages of the degradation little change in the bulk properties of the sample is discernible. Despite this fact subsequent events show that a gradual diffusion of water into the bulk occurs from the outset and is obviously accompanied by progressive chain scission within the polymer matrix.

The most obvious events, however, are those occurring at the polymer surface. An increase in surface energy reflects the increase in the concentration of hydroxyl and carboxyl groups at the surface as a consequence of ester hydrolysis at the polymer-water interface. As this process progresses the resultant surface erosion produces changes in surface rugosity that in turn produce changes in the goniophotometric scattering envelope. Initially, the changes in rugosity consist of an increasing concentration of irregularities that are smaller than the wavelength of light. These irregularities reduce the intensity of the scattered light without greatly affecting the dispersity of the scattering, and as a result, the value of *Is* drops without any concurrent increase in half-height peak width.

Table 1. Bioerodible polymers.

Polymer Name	Commercial Name	Structure
Poly(glycolic acid)	Dexon® (sutures)	$(\text{—O—CH}_2\text{—CO—})_n$
Poly((D,L-lactic acid)	—	Me $(\text{—O—C—CO—})_n$ H
Poly(glycolic-co-lactic acid)	Polyglactin 910 or Vicryl® (sutures)	Me $(\text{—O—CH}_2\text{—CO—O—C—CO—})_n$ H
Polyvalerolactone	—	$(\text{—O—(CH}_2)_4\text{—CO—})_n$
Poly(ϵ-caprolactone)	—	$(\text{—O—(CH}_2)_5\text{—CO—})_n$
Poly(ϵ-decalactone)	—	nBu $(\text{—O—C—(CH}_2)_4\text{—CO—})_n$ H
Poly(hydroxy butyrate)	Biopol	Me $(\text{—O—C—CH}_2\text{—CO—})_n$ H
Poly(hydroxy valerate)	(used as a comonomer with the butyrate)	Et $(\text{—O—C—CH}_2\text{—CO—})_n$ H
Polydioxanone	PDS (sutures)	$(\text{—O—(CH}_2)_2\text{—O—CH}_2\text{—CO—})_n$

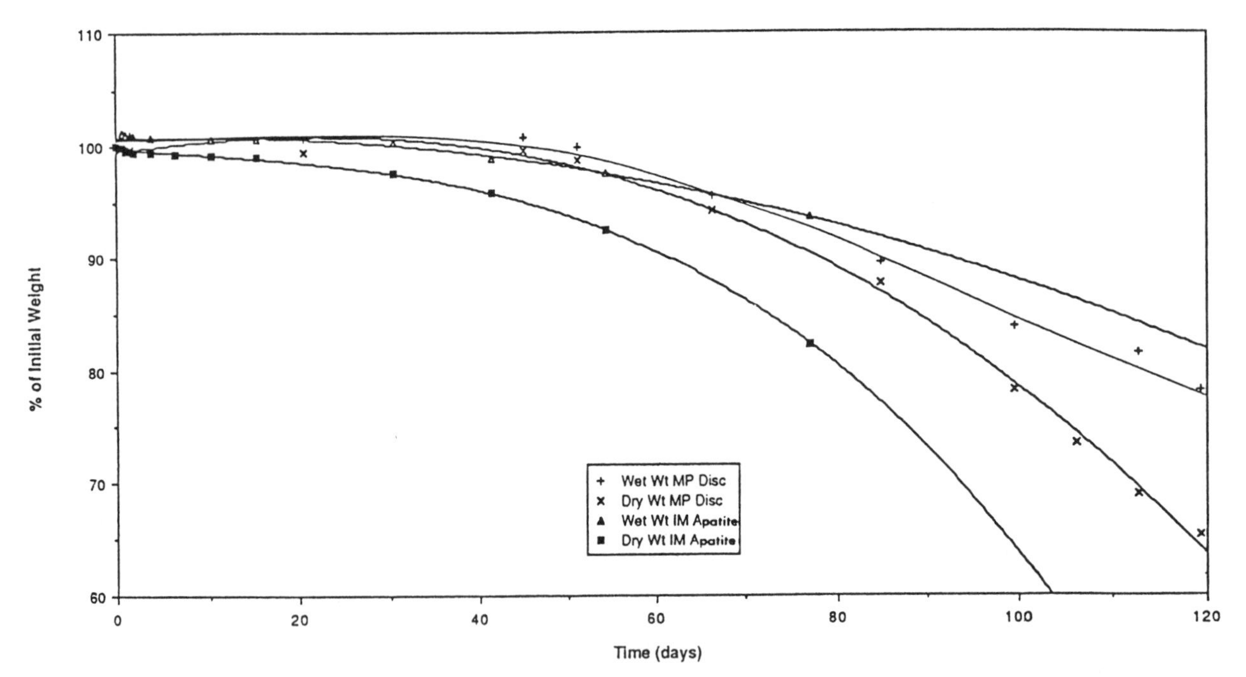

Figure 8. Comparison of progressive weight loss (measured wet and dry) of 12% PHV copolymer melt pressed discs and injection moulded plaque specimens at 70°C and pH 7.4.

The advancement of surface erosion leads to an increase in the size of the physical irregularities at the surface until they reach and exceed the wavelength of the incident light beam. This situation is illustrated in the difference between Figures 6 and 7, which have been referred to earlier. It is apparent that these progressive changes at the surface are occurring concurrently with a bulk erosional process that results from diffusion out of the matrix of the products of chain scission processes. Initially, only very low molecular weight fragments are able to diffuse out but as the process proceeds the matrix becomes more porous allowing an increasing loss of higher molecular weight degradation products. The increasing porosity of the matrix is reflected in the in-

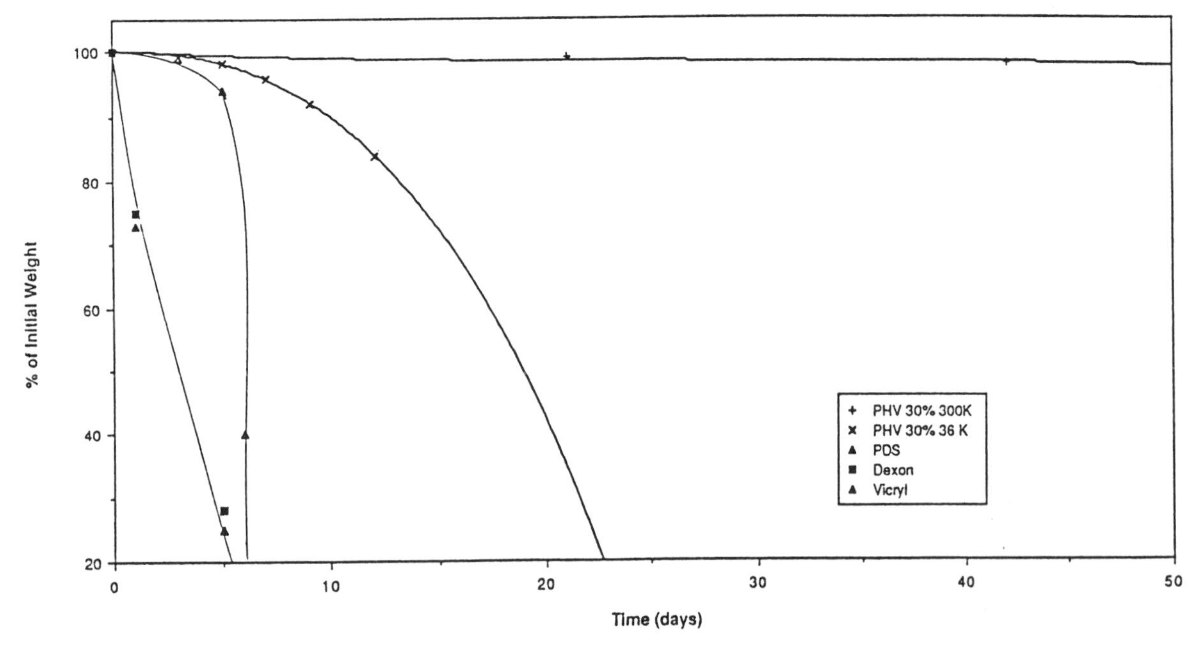

Figure 9. Comparison of progressive weight loss of 20% PHV copolymers with suture materials from Table 1 at 70°C and pH 7.4.

creasing difference between dry and wet weight of the samples, and in the almost autocatalytic profiles of the degradation curves. These are illustrated in Figures 8 and 9. Scanning electron microscopy shows that the erosional process occurs preferentially in amorphous regions, leaving a porous interconnected spherulitic morphology. Studies of this type involving a greater understanding of the effect of polymer structure on both the rates and mechanisms of degradation are of great value in both the previously mentioned applications, surgical fixation and drug delivery systems. In summary then, advancement of the surface erosional process leads to an increase in the size of surface irregularities until they reach and exceed the wavelength of the incident light beam. This situation produces the greater dispersity of the scattered light beam evident in Figure 7. Thus, goniophotometry provides an extremely sensitive method of monitoring the early stages of this, and any other polymer ageing process, in which surface degradation presages or parallels bulk degradation.

GAS PLASMA STUDIES

Another area in which the goniophotometer has given useful information involves the use of reactive and inert gas plasmas for modification and fabrication of polymer devices to be used at biological interfaces. The erosion appears (SEM) to produce surface rugosity in the 200 nm region. This rugosity produces no peak asymmetry at wavelengths generally used and work at shorter wavelengths is valuable. Similarly, plasma modification of polymer surfaces for use at biological interfaces is important in tissue culture and ocular spoilation studies [5].

All of these areas make use of the unusual features of the role of aromaticity in polymers in relation to their behaviour in oxygen plasmas. Although the applications are undoubtedly of some general interest, it is the unusual nature of the oxidation processes involved that will be of greatest relevance at this conference. Some of our findings are summarised below.

(a) In methyl methacrylate-styrene copolymers, relatively small amounts of styrene are required to produce a dramatic increase in polymer stability.

(b) The rate of polystyrene degradation in an oxygen plasma is independent of molecular weight between 3750 and 3,000,000 ($Mw/Mn = 1$).

(c) The position of pendant aromatic groups in carbon backbone polymers had little effect on the rate of plasma degradation.

(d) Interruption of potential zip sequences with hetero atoms had little effect on polymer stability.

(e) The role and positional independence of aromaticity in relation to the plasma degradation of a series of polyesters paralleled that observed in carbon backbone polymers.

Conventional atmospheric oxidation processes cannot represent the full, or even the major, explanation for this behaviour. The most probable explanation lies in the known ability of ground state oxygen atoms to react with aromatic rings in an addition reaction. This process removes oxygen atoms without forming free radicals and provides the basis for an interpretation of these observations [6]. This unusual behaviour of aromatic systems has led to the design of new membranes for transistor-based biosensors and to the development of highly active substrates for tissue culture [5].

ACKNOWLEDGEMENTS

The author wishes to express his appreciation for the contributions of co-authors in the listed references.

REFERENCES

1. Tighe, B. J. in *Polymer Surfaces*. D. T. Clark and W. J. Feast. Wiley, Chapter 14 (1978).
2. *E.S.B. Biomaterials Congress, Bologna, Sept. 14–17, 1986*. Abstracts; D. F. Williams (p. 192) and W. Bonfield (p. 193).
3. Holland, S. J., P. L. Gould and B. J. Tighe. *J. Controlled Release*, 4:155 (1986).
4. Holland, S. J., A. M. Jolly, M. Yasin and B. J. Tighe. *Biomaterials* (in press).
5. Thomas, K. D., M. J. Lydon and B. J. Tighe in *Biological and Biomechanical Performance of Materials*. P. Christel, A. Meunier and A. J. C. Lee, eds. Elsevier, p. 379 (1986).
6. Moss, S. J., A. M. Jolly and B. J. Tighe. *Plasma Chem. Plasma Processing*, 6:401 (1986).

A. TKÁČ[1]

ESR Investigation of Spontaneous and Controlled H-Transfer and Electron-Transfer Mechanisms in Oxidation

ABSTRACT

Some common aspects of atomic hydrogen transfer and of one electron transfer mechanisms carrying the spontaneous and controlled free radical oxidation on model systems in gaseous, liquid, molten and solid phases applying four different ESR techniques are discussed:

(a) The direct indication of atomic hydrogen and HO· radicals in an H_2/O_2 flame delivers quantitative data for deeper understanding of the chain branching explosive cycle and of the flame oscillation.
(b) A universal scheme of H-transfer reactions of burning based on identification of free radicals extracted from different zones of the flame and of radicals remaining in the residue after vacuum pyrolysis of polymers was elaborated.
(c) The higher H-transfer efficiency of the multifunctional phenolic antioxidant Ethanox in comparison with Irganox was tested by coordinated peroxy radicals. In a mixture of the two phenolic antioxidants an H-transfer from Ethanox to the less stable radicals generated from Irganox proceeds as long as equilibrium between the two phenoxy radicals is reached.
(d) The elementary steps of the redox cycle of Ni(I)–Ni(II)–Ni(III) in competition with electron transfer mechanisms of Fe(III)–Fe(II)–Fe(III) controlling the programmed antioxidant and post-prooxidant efficiency in the presence of hydroperoxides and antioxidants in mixtures with different molar ratios of Ni(II) dipropyldithiocarbamate and Fe(III) dipropyldithiocarbamate is described.

KEY WORDS

ESR, free radicals, flame retardation, phenolic antioxidants, dithiocarbamate metal chelates.

INTRODUCTION

Two elementary steps, *electron-transfer* (ET) and *hydrogen-transfer* (HT) control the oxidation mechanism on a molecular level. One of the apparently simple but fundamental reactions between hydrogen and oxygen—the formation of water molecules

$$2H_2 + O_2 \rightarrow 2H_2O$$

involving, simultaneously, transfer of four electrons and four H-atoms—can be discussed from two extremely different points of view:

(a) As a spontaneous continuously branched free radical reaction at explosion of a stoichiometric H_2/O_2 mixture with irreversible heat release
(b) As a carefully, through metal enzymes, controlled process on cell level at physiological temperature, in which the HT steps are separated from the ET steps by means of biological membranes, where a part of the free energy remains chemically stored

In practical life we are daily confronted with oxidative mechanisms operating between these two extremes: spontaneous and controlled ET and HT steps. Consider the polymers, e.g., at ambient temperature, exposed to slow random oxidation, to aging, or at elevated temperature they can start to burn, often causing uncontrolled fire.

A present-day problem of polymer technology is not only to slow down the rate of oxidation by applying antioxidants and peroxide decomposers, or lowering the flammability using flame retardants—but also to prepare systems with time controlled programmed stability coupled with a catalytically accelerated oxidative decomposition after a timed service life.

A useful experimental technique giving quantitative and qualitative information of free radical reactions in consequence of HT or ET mechanism, when transition metals are operating, is electron spin resonance (ESR).

The aim of this lecture is to present some ESR methods applied in our laboratory for studying HT and ET mechanisms in gaseous, liquid and solid phase.

The Direct Indication of H· and HO· Radicals in an H_2/O_2 Flame

To situate a well-defined flame directly in a resonance cavity of an ESR spectrometer presents many experimen-

[1]Institute of Physical Chemistry, Faculty of Chemical Technology, Slovak Technical University, 812 37 Bratislava, Czechoslovakia.

tal difficulties. Although the temperature of the flame can reach more than 1000°C, the cavity must remain rigorously at ambient temperature. As it was shown, the use of a quartz evacuated Dewar cell for thermal isolation is insufficient, but we were successful in applying a quartz double jacket cell with effective cooling by means of compressed cooled air or nitrogen (Figure 1). Pure hydrogen or a mixture of H_2/O_2 in the desired molar ratio passes through a quartz capillary burner axially situated in the cell. The upper end of the cell is connected to a vacuum line and a cooled trap, in which the product of burning H_2O can be quantitatively frozen out.

On the bottom of the cell is a valve which regulates the intake of air and keeps a constant diffuse H_2 flame burning. When oxygen is added, the diffused flame can be stepwise converted to a premixed laminar flame. The burner is movable which allows the flame to be situated at different heights in the cavity.

The dominant ESR signal identified in this way is composed of two narrow lines with a coupling constant a = 50.5 mT and an effective g-value of g_{ef} = 2.014, which means that the center of the symmetry between the two lines is shifted by 2 mT to the lower side of the applied homogeneous magnetic field B in comparison to the g-value of free electron 2.0028. This signal is characteristic for atomic hydrogen with the free electron in an s-orbital (S = 1/2) being in strong 100% contact interaction with the proton of the nucleus possessing a nuclear magnetic moment of I = 1/2 (Figure 2). The experimentally measured parameters of the ESR signal are in perfect correlation with the theoretically calculated ones from the

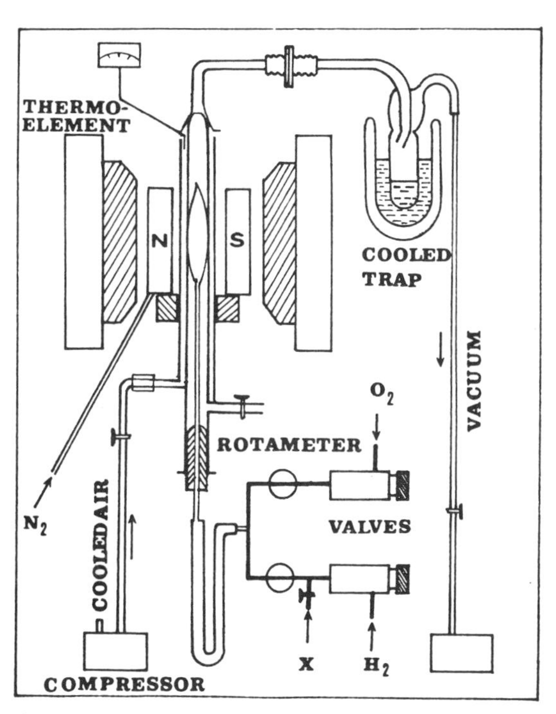

Figure 1. Cooled quartz cell with the flame in the ESR cavity.

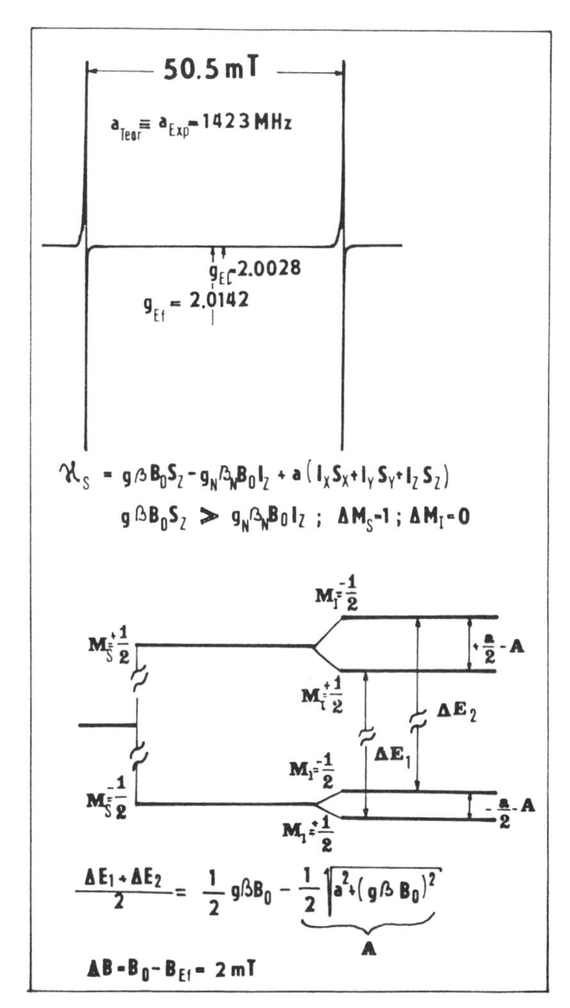

Figure 2. ESR signal of H· in the H_2/O_2 flame and the energy diagram resulting from solution of the spin Hamiltonian.

spin Hamiltonian $\mathcal{H}_S$, taking into consideration an isotropic solution in the second order and a projection to the z-axis parallel to the applied magnetic field B. In the Hamiltonian g and g_N are the g-values of the free electron and of the nucleus, S and I are the electron and the nuclear spin vectors and a the hyperfine splitting constant. From a quantum chemical point of view, the s-electron (which is in contact with the proton) has a probability electron density $|\Psi(O)|^2$ equal to 1—so that the theoretical coupling constant of the atomic hydrogen can be evaluated from the Fermi-equation [1]:

$$a = \frac{8\pi}{3} g\beta g_N\beta_N |\Psi(O)|^2 = 1423 \text{ MHz} = 50.5 \text{ mT}$$

which is in agreement with the experimentally observed a = 50.5 mT of H· in the flame. Similarly the shift of the center of the signal symmetry from the position of the g-value of free electron g = 2.0028 is in very good correlation with the theoretically calculated values.

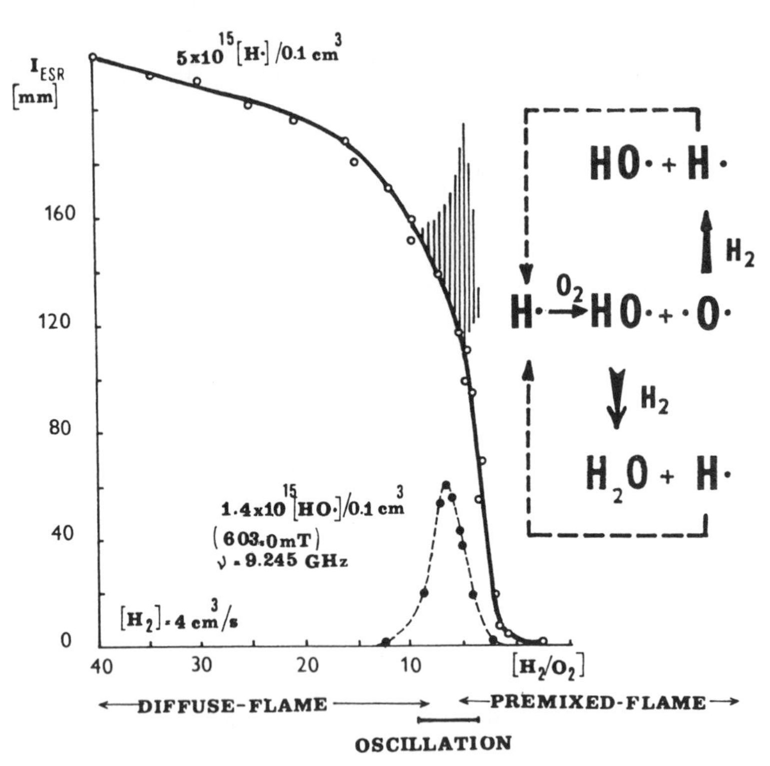

Figure 3. Dependence of the H· ESR signal height upon the molar ratio of the H_2/O_2 in the premixed gases led into the burner. The flowing rate of H_2 was constant 4 cm^3s^{-1}.

The concentration of H· in the diffuse flame, when pure hydrogen burns with air drawn in by a weak vacuum, is surprisingly high (on the order of 6–7 $\times$ 10^{15} in 0.1 cm^3). At a constant flow rate of H_2 (4 cm^3 $\times$ s^{-1}), the concentration of H· successively decreases, when oxygen is stepwise admixed into the hydrogen stream. From the approximate molar ratio of H_2:O_2 = 10:1, a rapid decrease in concentration of atomic hydrogen takes place. Just before reaching the stoichiometric ratio 2:1, the concentration of H· falls under the threshold level of sensitivity of the ESR spectrometer (3 $\times$ 10^{12} spins/0.1 cm^3). Between the ratio 7:1 and 3:1 a qualitative change of the flame mechanism is observed. The H· concentration together with the flame height begins to oscillate reaching a maximum of 50% change of the ESR signal intensity at the ratio 4.7:1 (Figure 3).

When the diffuse character of the flame is changed to the premixed flame (H_2:O_2 = 2:1), the ESR signal of H· disappears. In the critical zone of the rapid decrease of atomic hydrogen, the concentration of HO· radicals (g = 1.1 at ν = 9.246 GHz) passes through a maximum. The expected presence of atomic oxygen ·O· as the product of the fundamental branching reaction of the explosive cycle (see Figure 3) could not be registered according to its high reactivity and very low actual concentration. From the reaction scheme it is clear that one H· in the presence of O_2 molecules can create three radicals, which during the continuous propagation branching cycle can lead to an explosive spontaneous HT.

In a flame fed with the fuel as fast as it reacts, while the products of burning are free to escape, we are dealing with a controlled explosion, where the steady state concentration of H· is 6 $\times$ 10^{15} spin/0.1 cm^3 at the temperature of about 450°C in a distance of 5 mm above the top of the burner. During this process about 10^{19} molecules of H_2O are formed in one second, so that we must expect at least 10^5 long kinetic chains. At the steady state the rate of initiation plus the rate of branching is equal to the rate of termination. When the admixing of O_2 into the H_2 diffuse flame, the degenerative chain branching is removed and the lifetime of hot radicals is shortened, the concentration of H· is no longer measurable.

Different kinetic chain lengths in the outer surface layer being in contact with the air and in the inner volume of the premixed flame can explain the observed flame oscillations. The described ESR technique can be advantageously used for studying the efficiency of chain brakers and flame retardants by admixing them into the H_2 stream in front of the burner.

Free-Radical Transfer From the Molten Into the Gaseous Phase of Thermally Treated Polymers and Flame Retardation

The retardation of polymer burning is a complex chemical and physical problem, the mechanism of which varies considerably in the individual zones of burning depending on the rate of oxygen diffusion and heat dissipation. By ESR, in combination with trapping of volatilized products of heated samples on a rotating cryostat cooled continuously by flowing liquid nitrogen, the concentration of the free radicals in the gaseous phase and in the molten polymer could be measured.

When the polymer under exclusion of oxygen is thermally decomposed in a vacuum of 6×10^{-2} Pa in a temperature range of 220–400°C, using a glass pyrolysis cell closed with a glass capillary jet situated about 2 mm under the cooled rotating drum, the concentration of highly reactive primary radical fragments after their immobilization can be determined by ESR. Consecutive contact of the frozen sample with oxygen enables the study of the elementary reaction of peroxy radical formation:

$$R\cdot + O_2 \rightarrow RO_2^{\cdot}$$

according to the transformation of the original symmetric singlet of alkyl radicals $g = 2.003$, to the characteristic asymmetric doublet signal of $RO_2^{\cdot}$ with $g = 2.009$–2.0145. In the presence of flame retardants or radical scavengers of the secondary less reactive paramagnetic species, the kinetics of their tranformation by stepwise temperature increase can be followed. For instance for a polypropylene sample (PP) containing a phosphor-brominated flame retardant (6% of Sandoflam), the concentration of the primary alkyl radicals transferred from the molten to the gaseous phase is reduced by two orders in comparison with PP without additive (from 3.5×10^{15} to 3.5×10^{13} spins/g), whereas, the limiting oxygen index (LOI) is increased from 17–18% to 25–26%.

In the residue after a 45 min long pyrolysis of different polymers the amount of the non-volatilized oligomeric and macroradical fragments can be also measured. The effective production of the gases (the fuel) enriched with reactive free radicals starts in different thermal regions according to the strength of the polymer bonds. The actual pressure above the polymer surface determines the temperature of inflammation of the heated polymer. The concentration ratio between radicals transferred from the molten to the gaseous phase and radical fragments having higher molecular weight than about 500 D formed by stepwise thermal decomposition of the polymer, when 4–6% of the original weight was volatilized, is compared in Figure 4 for the different polymers.

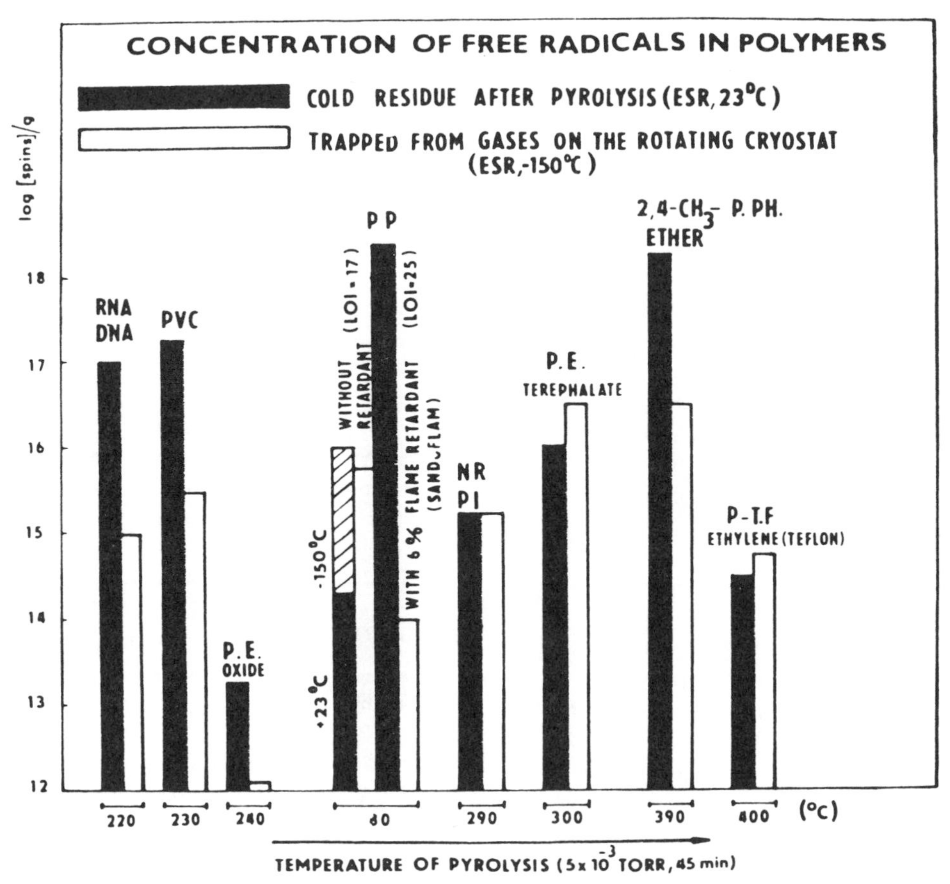

Figure 4. Concentration of radicals trapped from the gaseous phase and in the molten residue after vacuum pyrolysis of different polymers.

At the same vacuum of 5×10^{-1} Pa the decomposition of the phosphoresteric bond of DNA and RNA proceeds at the lowest temperature range 200–220°C; at 230°C the dehalogenation of polyvinylchloride begins forming high concentrations of polyenic radicals in the residue. The presence of 6% phosphorbromic retardant (Sandoflam) in polypropylene changes dramatically the ratio of radical concentration in the gaseous and in the molten phase. So far in the gaseous phase the radical concentration falls by about two orders (10^{14} spin/g), in the molten phase increases about by three orders (10^{18} spin/g). The creation of polyenic radicals in the preflame zone is catalyzed by the evolution of bromine and the simultaneous formation of HBr after an HT reaction deactivates the $RO_2^{\cdot}$ in the surface, while the HO· radicals in the gaseous phase are under control of the atomic bromine

$$RO_2^{\cdot} + HBr = RO_2H + Br\cdot$$

$$HO_2^{\cdot} + Br\cdot = HBr + O_2$$

When gases from the dim, high-temperature, pyrolytic zone of a burning polymer (or that of a burning wax candle) are led to the surface of the rotating cryostat by means of the thermally isolated quartz tube, there is (in the condensed white explosive vapour) a high concentration of reactive radicals (0.2–0.4 mT broad ESR single line $g = 2.0028$) obtained which irreversibly disappears on heating. In contrast, when the gases are extracted from the light-emitting zone of the flame (with the glowing carbon particles), black soot condenses on the rotating drum giving a single sharp ESR line of polyaromatic or graphitic radicals which are perfectly stable at ambient temperature.

A General Scheme of Radical Reactions in the Oxygen Containing and in the Oxygen Free Zone of a Diffuse Flame

A universal scheme (Figure 5) of radical reactions in the flame can be proposed by comparing the ESR studies of the spontaneous HT reactions directly in the gaseous phase of a diffuse and of a laminar premixed flame of H_2/air or H_2/O_2 as well as of the feature of radicals extracted from different zones of the diffuse flame of a burning candle or polymer sample.

In the first column of Figure 5 the main radical reactions running in the thin outer zone of the diffuse flame

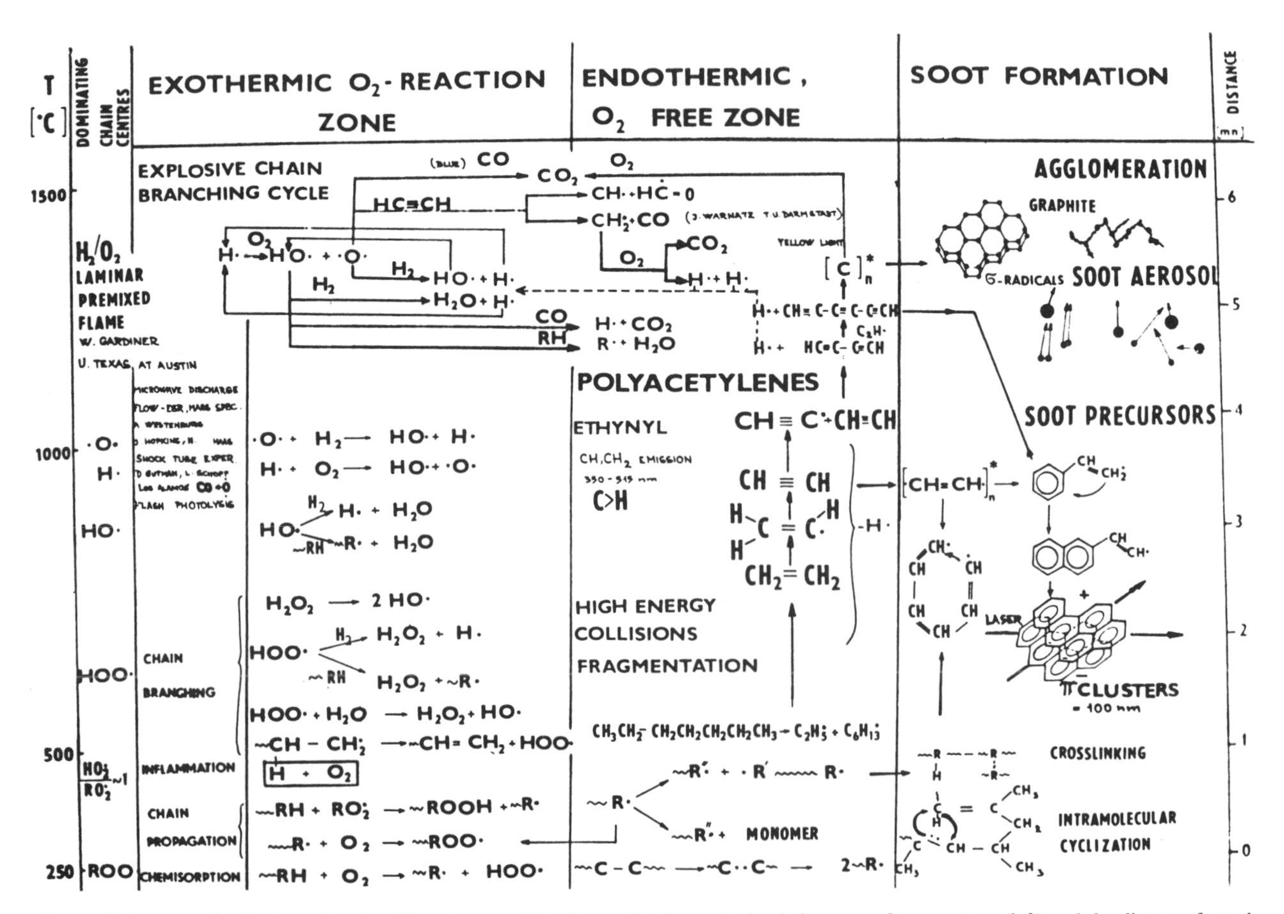

Figure 5. Scheme of radical reactions in different zones of the flame. On the vertical axis increase of temperature (left) and the distance from the top of the burner (right).

in contact with oxygen (or in the premixed flame fuel/oxygen) are summarized in a vertical arrangement of temperature increase and of distance from the top of the burner. In the middle column are registered, the endothermic reactions of thermal decomposition in exclusion of oxygen, where the stepwise shortening of volatilized molecules proceed, followed by dehydrogenation and so leading to more and more thermally stable structures with double and triple bonds. The elimination of hydrogen atoms connected with creation of the ethynyl radical $C_2H\cdot$ triggers the polyacetylene, carbon fibres and polyaromatic molecule formation. The collision of alkyl radicals carrying high kinetic energy, leads to secondary dehydrogenation and carbene production:

$$CH_3 \rightarrow :CH_2 + H\cdot$$

Carbenes in contact with molecular oxygen generate aldehydes and atomic oxygen:

$$:CHR + O_2 = RCHO + \cdot O\cdot$$

In the third column there are indicated chemical and physical processes leading to soot formation as well as to synthesis of toxic organic polyaromates after cyclization (e.g., the carcinogenic 3,4-benzopyrene).

The inactivation of the chain-carrying radicals in the gaseous phase not only lowers the heat transfer into the preflame zone, but also shifts the reaction mechanism in the scheme to soot formation. The programmed simultaneous volatilization of flame retardants together with the production of reactive radical fragments from the thermally decomposed polymers is an essential step in controlling the temperature of flammability.

HT in Multicomponent, Multifunctional Antioxidant Systems

The thermal [3] and photochemical [4] oxidative polymer degradation evolves slowly with time to obey the kinetic equations of radical reactions with degenerative chain branching [5]. The most important chain branching step is the homolytic decomposition of hydroperoxides:

$$ROOH \rightarrow RO\cdot + HO\cdot$$

Deactivation of primary radicals by HT creating secondary radicals with prolonged mean lifetime is a fundamental step for interruption of the chain propagation in the presence of antioxidants:

$$RO\cdot + AH \rightarrow ROH + A\cdot$$

The usefulness of the ESR technique in elucidation of relations between the structure and reactivity of radicals generated from antioxidants during HT and ET processes will be demonstrated on two commercial, multifunctional, phenolic antioxidants: Ethanox 330 (1,3,5-trimethyl-2,4,6-tris(3,5-di-tert-butyl-4-hydroxy-benzyl) benzene), possessing three functional OH groups, and on Irganox 1010 (tetrakis-3-(3,5-di-tert-butyl-4-hydroxy phenyl)propionyl oxymethyl methane) possessing four OH groups.

The results are based on HT reactions realized at ambient temperature between tert. butyl peroxy radicals stabilized by coordination on acetylacetonate of cobalt $Co(III)RO_2^{\cdot}$ [6] and the antioxidant dissolved in nonpolar solvents.

When H-atom is abstracted by $RO_2^{\cdot}$ radicals from Ethanox dissolved in benzene, an additive ESR spectrum (A + B) of two phenoxy radicals with the same g-value 2.0039 is obtained (Figure 6). According to computer simulation, this signal was interpreted as 60% of signal A having two equivalent hydrogens of the $-CH_2-$ bridge in para-position [$2 \times a(CH_2) = 1.333$ mT] and two equivalent meta-hydrogens ($2 \times a_m = 0.162$ mT); plus 40% of signal B with unequivalent H-atoms in the methylene bridge [$a^1(CH_2) = 1.267$ mT, $a^2(CH_2) = 1.397$ mT] and two equivalent meta-hydrogens ($2 \times a_m = 0.162$ mT).

Both the radicals having relatively high spin density in para-positions differ in the mobility of the hydrogens in the $-CH_2-$ bridge. In the type A the two hydrogens can freely rotate lying symmetrically in the π-conjugated aromatic plane, while in the type B (according to steric hindrance) the two H-atoms of the methylene bridge are not equivalent.

Quantitative measurements have shown that the shape and the intensity of the additive (A + B) ESR signal is a sensitive function of the absolute concentration of the initiating $Co(III)RO_2^{\cdot}$ radicals. When the concentration of Ethanox is in the range of the initiating radicals (10^{-4} M), only the original broad signal of coordinated peroxy radicals with $g = 2.0145$ disappears without the generation of equivalent amounts of secondary phenoxy radicals from the antioxidant. The AO· and $RO_2^{\cdot}$ recombine after coupling in the para-position of the resonance cyclohexadienonyl structure

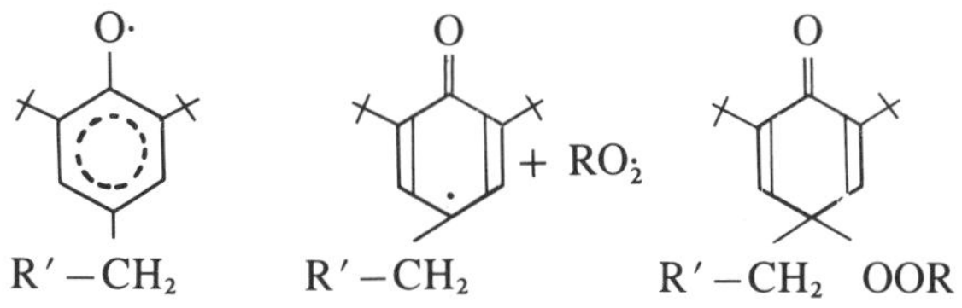

O· O O

+ $RO_2^{\cdot}$

$R'-CH_2$ $R'-CH_2$ $R'-CH_2$ OOR

When the radical attack results in a great surplus of Ethanox relative to the concentration of coordinated peroxy radicals (2.5×10^{-1} M $\gg 10^{-5}$ M), an equivalent concentration of the A-type phenoxy radical is gener-

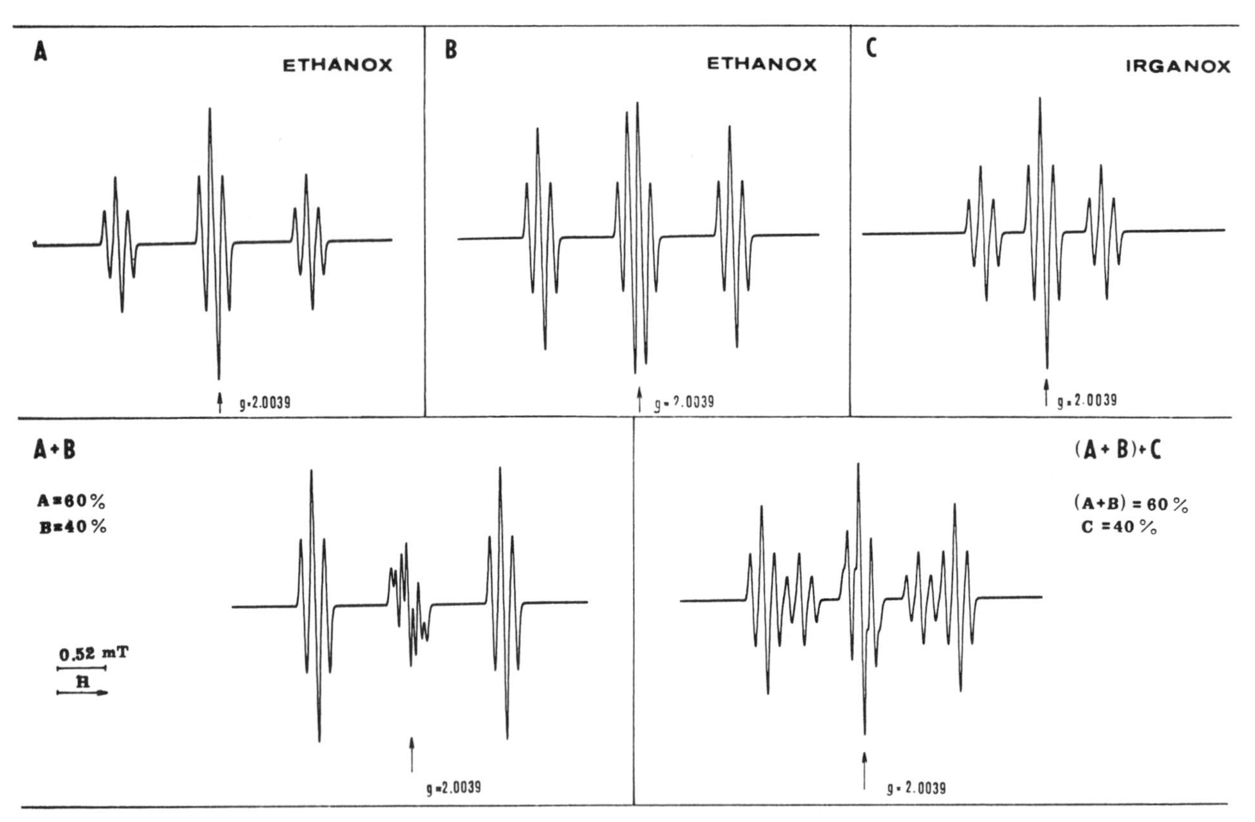

Figure 6. (A + B) experimental ESR spectrum of two phenoxy radicals generated with coordinated peroxy radicals Co(III)$RO_2^{\cdot}$ in 0.25 M benzene solution of Ethanox. Individual signals for spectral simulation A (equivalent hydrogens in the methylene bridge), B (non-equivalent H-atoms of the ethylene bridge). ESR of the phenoxy radical generated from Irganox (C). Additive ESR signal A + B + C generated in an equimolar solution of Ethanox and Irganox.

ated. In contrast, when a product of the Ethanox oxidation by tert.butyl hydroperoxide (having two cyclohexadienone and one remaining OH group) is attacked by $RO_2^{\cdot}$ radicals, only the B type ESR signal, characteristic of two unequivalent hydrogens of the methylene bridge, is formed. The relatively high coupling constant in para-position [$a(CH_2)$ = 1.2–1.3 mT] is the result of a strong pull effect of the central, methylated aromatic ring on the unpaired electron, which is in contrast to phenoxy radicals generated from Irganox where the delocalization in the para-position is at least 20% lower [$2 \times a(CH_2) = 0.828$ mT, $2 \times a_m = 0.168$ mT, Figure 6C).

Taking all these facts into consideration, the additive (A + B) ESR signal generated from Ethanox is the result of a multi-step radical attack (see below).

The lower stability of radicals generated from Irganox in comparison with phenoxy radicals of Ethanox was documented also according to a secondary HT in the mixture of the two antioxidants. When a high concentration of phenoxy radicals from Irganox is generated and

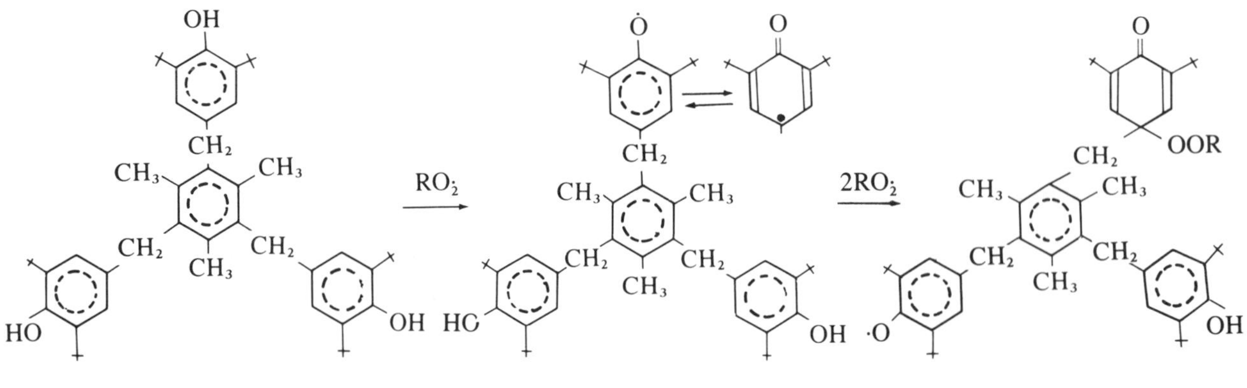

$\xrightarrow{RO_2^{\cdot}}$ $\xrightarrow{2RO_2^{\cdot}}$

then an equivalent amount of Ethanox is added, the original 3-line signal of phenoxy radicals effectively decreases, and in the final superimposed ESR signal the intensity of the lines belonging to phenoxy radicals of Ethanox prevails in a relation 2:1. But attacking an equimolar 1:1 mixture of the two antioxidants (0.25 M, benzene) with coordinated $RO_2^{\cdot}$, an additive ESR signal composed of three individual radical types A 36%, B 24% and C 40% is observed. According to the experiments presented, an HT from the more efficient antioxidant Ethanox to the less stable phenoxy radicals of Irganox takes place up to the following equilibrium:

$$R_2-CH_2-CH_2-C_6H_4-O\cdot \; + \; HO-C_6H_4-CH_2-R' \rightarrow$$

Irganox Ethanox

$$R-CH_2-CH_2-C_6H_4-OH \; + \cdot O-C_6H_4-CH_2-R'$$

A similar but multi-step HT has been proved also in biological systems, when the radical initiation was started with peroxy radicals coordinated on hemoglobin or on different ferrihemoproteins [7]. The most stable radicals of biological antioxidants (α-tocopherol, ascorbic acid, glutathione) are situated on the end of such HT cascades at physiological conditions. Radicals with intermediate reactivity generated from arylamines and azoarylamines as H-donor antioxidants, often with carcinogenic properties, lie in the middle of this HT-chain between the highly reactive oxygen radicals $HO\cdot$, $HO_2^{\cdot}$, $O_2^{\cdot}$ and the low reactive radicals of biological antioxidants.

Chelated Transition Metals as Peroxide Decomposers and ET in Programmed Anti- and Prooxidant Systems

It is known that phenols and aromatic amines begin to lose their HT antioxidant efficiency at temperatures over 100°C, because of the homolytic decomposition of the peroxides formed and direct reaction with oxygen. To overcome this disadvantage a second class of preventive antioxidants has been developed which reduces the rate of chain initiation by ET mechanism often combined with ligand transfer (LT). Organic compounds containing at least one sulphur atom, are the most important peroxide decomposers and more potent radical scavengers. Coordination compounds of Ni(II), Zn(II), Fe(III) with chelated dithiocarbamate ligands (DTC):

$$R_2N-C\langle{}^{S}_{S}\rangle M\langle{}^{S}_{S}\rangle C-NR_2$$

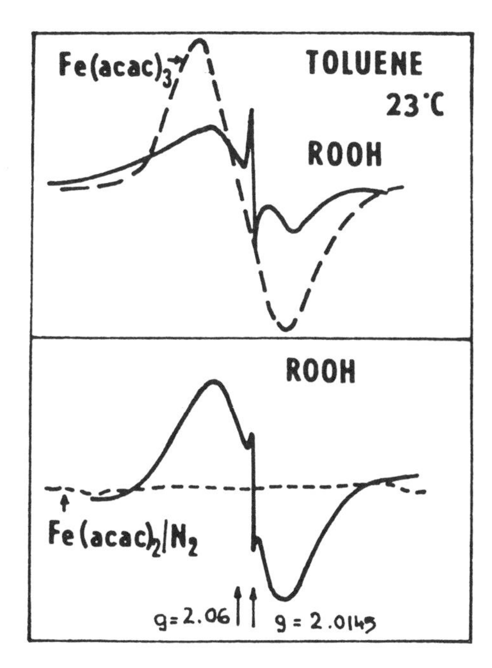

Figure 7. ESR of 0.05 M toluene solution of $Fe(acac)_3$ and of $Fe(acac)_2$ (dotted line) and changes of the signal after reacting with ROOH.

are an important group of preventive antioxidants. But according to the experimental conditions, they can also demonstrate prooxidant effects by producing free radicals during peroxide decomposition [9].

A comprehensive study of $Ni(DTC)_2$ as photoantioxidant with $Fe(DTC)_3$ as photoprooxidant lead S. AL-Malaika, A. Margoli and G. Scott [10] to describe an effective photo-antiprooxidant system for polyethylene, in which the induction period is controlled by the nickel complex and the post-induction rate by the iron complex.

Our systematic ESR study gave us more detailed information on the elementary ET, HT and LT steps in such multicomponent system controlling the time-variable thermal or photo-stability.

Let us start with iron(III)acetylacetonate $Fe(acac)_3$, which is known to be the highly active photoactivator for polyethylene, but suffers from the practical deficiency that it is also a powerful prooxidant for polyolefins during processing [11]. In the presence of tert.butyl hydroperoxide (ROOH) the radical chemistry of $Fe(acac)_3$ is quite different from that of dipropyledithiocarbamate of iron $Fe(DTC)_3$.

According to the multi-step ET mechanism between $Fe(acac)_3$ dissolved in toluene and ROOH (in a ten-fold molar surplus), the original broad ESR signal of the low-spin $3d^5$ Fe(III), $g = 2.06$, decreases and is superimposed by the signal of coordinated $RO_2^{\cdot}$ radicals on Fe(IV) characterized by $g = 2.0145$ (Figure 7). $Fe(acac)_2$ at the same conditions, originally without ESR signal ($3d^6$) is oxidized to Fe(III) and a relatively high

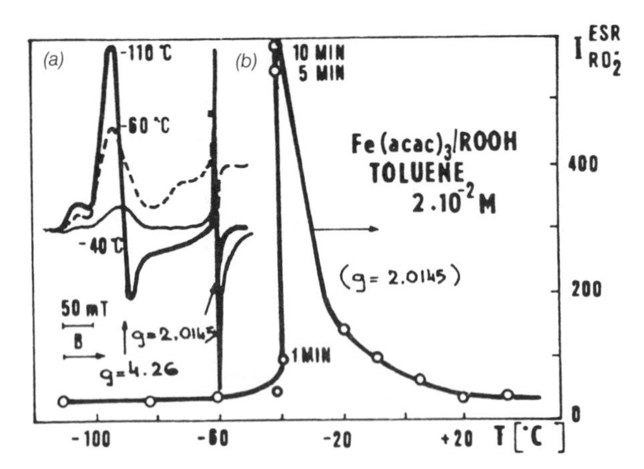

Figure 8. (a) Decrease of the high-spin ESR signal of Fe(III) in 0.02 M toluene solution of $Fe(acac)_3$ and (b) increase of the signal of peroxy radicals in the course of heating a rapidly freezed mixture with ROOH surplus.

concentration of coordinated Fe(IV) $RO_2^{\cdot}$ radicals is formed.

$$FeL_3 + ROOH \rightarrow HOFeL_2 + ROL$$

$$HOFeL_2 + 2ROOH \rightarrow L_2Fe\overset{\overset{\displaystyle R}{|}}{\underset{\underset{\displaystyle O}{\|}}{O}}{-}O\cdot + ROH + H_2O$$

The direct ET to ROOH from the completely liganded iron(III) resulting in generation of "free," together with "coordinated," tert.buryl peroxy radicals was proved in

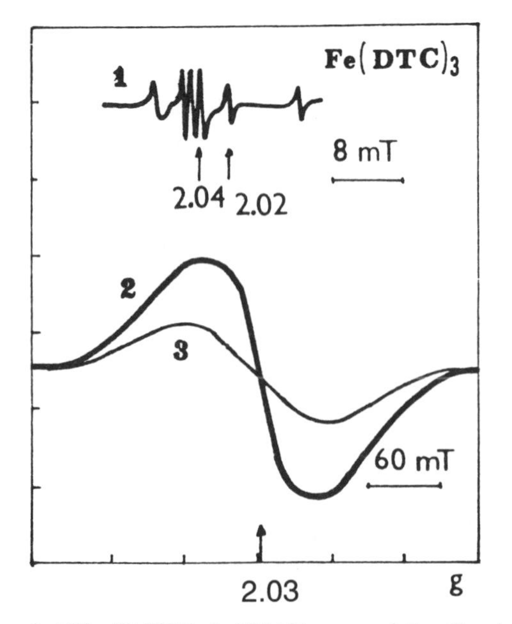

Figure 9. ESR of $Fe(DTC)_3$ in 0.05 M benzene solution (1) and after oxidation with 0.16 M (2) and 0.32 M (3) ROOH.

the course of rapid freezing to $-20°C$ of a 2×10^{-2} M toluene solution of $Fe(acac)_3$ (Figure 8). The electron transfer starts in the thermal interval, where the frozen solid phase changes to liquid one, near the temperature of $-40°C$ accompanied with an effective increase of free $RO_2^{\cdot}$ radicals. Passing a narrow maximum the concentration of free $RO_2^{\cdot}$ rapidly decreases as temperature increases. At ambient temperature, at least 2–5% of all generated peroxy radicals remain stabilized on the highest oxidation state of Fe(IV). The temperature range in which the free $RO_2^{\cdot}$ are generated, is the same in which the reversible change of high spin Fe(III), $S = 5/2$, $g = 4.26$ to low spin Fe(III), $S = 1/2$, $g = 2.06$ is observed. High-spin forms of iron cannot be identified by means of ESR in liquid phase in consequence of their short relaxation times. This is also valid for $Fe(DTC)_3$ already at ambient temperature so that in a 0.05 M benzene solution only the signal of the free dithiocarbamate ligand radicals L· can be seen (Figure 9). The ESR signal is composed of two triplets with different g-values and different coupling constants of the free electron with the nitrogen nucleus ($I = 1$) of the ligand. The high g-value 2.02 with the greater coupling constant $a_N = 8$ mT is characteristic of the sulphuric ligand radical.

$$\left[\begin{matrix} C_3H_7 \\ C_3H_7 \end{matrix} \right\rangle N{-}C \left\langle\begin{matrix} S \\ \bullet \\ S \end{matrix}\right]$$

The interpretation of the second triplet signal with the highest g-value 2.04 and lower $a_N = 1.4$ mT is for the present not clear and we suggest an internal ET of $Fe(DTC)_2$.

$$Fe(DTC)_3 \rightarrow Fe(DTC)_2 + (DTC)^{\cdot}$$
$$(g = 2.02,\ a_N = 8\ \text{mT})$$

$$Fe(DTC)_2 \rightarrow [Fe^{+}DTC^{\cdot-}(DTC)]$$
$$(g = 2.04,\ a_N = 1.4\ \text{mT})$$

The ESR signal of the high spin Fe(III) $3d^5$ appears ($g = 2.03$) only after reaction of $Fe(DTC)_3$ with ROOH and after disappearance of both ligand radicals and oxidative destruction of the chelated complex. The presence of antioxidant slows down the catalytic decomposition of $Fe(DTC)_3$ during reaction with ROOH by scavenging free $RO_2^{\cdot}$ radicals (Figure 10).

$$Fe^{III}L_3 + ROOH \rightarrow Fe^{II}L_2 + LH + RO_2^{\cdot}$$

$$Fe^{II}L_2 + RO_2^{\cdot} \rightarrow L_2Fe^{III}OOR \rightarrow L_2Fe^{III}OR + \tfrac{1}{2}O_2$$

$$RO_2^{\cdot} + AH \rightarrow ROOH + A\cdot \qquad (g = 2.0039)$$

The same type of sulphur ligand radical L· ($g = 2.02$,

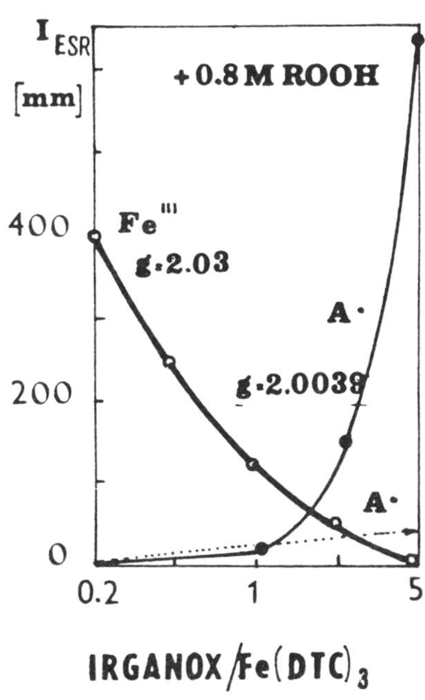

Figure 10. Dependence of the low spin Fe(III) signal generated in oxidation of $Fe(DTC)_3$ with 0.8 M ROOH, and of the simultaneous increase of the free phenoxy radical A· of Irganox, upon the molar ratio Irganox/$Fe(DTC)_3$. Increase of A· without ROOH (dotted line).

a_N = 8 mT) seen at $Fe(DTC)_3$ is observed in the benzene solution of $Ni(DTC)_2$, but together with the signal of the reduced nickel $3d^9$ Ni(I)L, g = 2.099 (Figure 11). The ESR signal of Ni(II), $3d^8$ in its high-spin state at ambient temperature is not observable. Just in the presence of traces of ROOH the ESR signal of Ni(I) is immediately

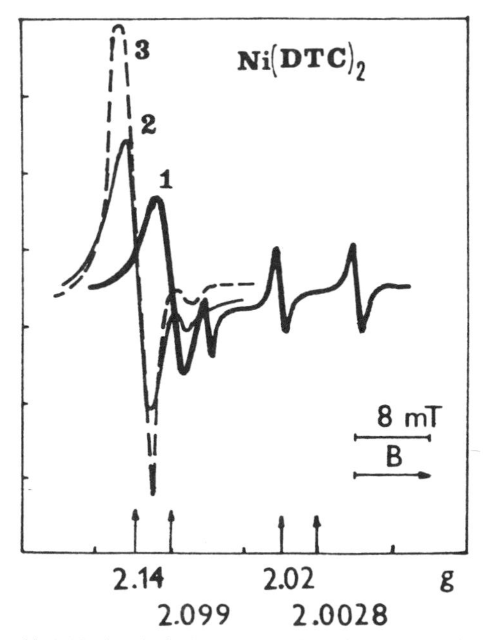

Figure 11. ESR signal of $Ni(DTC)_2$ in 0.05 M benzene solution (1) and after oxidation with 0.16 M (2) and 0.32 M (3) ROOH.

transformed to a new single line shifted to higher g = 2.14 as the result of nickel oxidation to Ni(III), $3d^7$ state

$$Ni^{II}L_2 \rightarrow Ni^{I}L + L\cdot$$
$$(g = 2.099, g = 2.02)$$

$$Ni^{II}L_2 + ROOH \rightarrow Ni^{III}L_2OH + RO\cdot$$
$$(g = 2.14)$$

$$Ni^{II}L_2 + RO\cdot \rightarrow Ni^{III}L_2OR$$
$$(g = 2.14)$$

At the molar ratio of 2:1 = $Ni(II)L_2$:ROOH a perfect control of hydroperoxide decomposition with simultaneous elimination of free radicals proceeds. But with increase of ROOH concentration up to a ratio 2:2, all ESR signals of nickel disappear:

$$HONi^{III}L_2 + ROOH \rightarrow HONi^{II}L + LH + RO_2^{\cdot}$$
$$(2LH + 2RO_2^{\cdot} \rightarrow 2ROH + L - L)$$

$$HONi^{II}L + ROOH \rightarrow HONi^{III}LOH + RO\cdot$$
$$\downarrow \qquad \searrow$$
$$Ni^{II}(OH)_2 + L\cdot \qquad (LOR)$$

The final reaction product of Ni(II) without (DTC) ligands cannot keep the peroxide decomposition under control simultaneously with radical scavenging by LT mechanism.

In a binary mixture of $Ni(DTC)_2$ and $Fe(DTC)_3$ (0.05 M benzene solvents) the original Ni(I)L signal at g = 2.099 effectively increases, which could be explained by an internal catalyzed ET redox cycle (Figure 12).

$$Fe^{III}L_3 \rightarrow Fe^{II}L_2 + L\cdot$$

$$Fe^{II}L_2 + Ni^{II}L_2 \rightarrow Fe^{III}L_3 + Ni^{I}L$$
$$(g = 2.099)$$

Already at a low concentration of ROOH the signal of Ni(I)L further increases, as a consequence of an ET from ROOH to $Fe(III)L_3$, producing $Fe(II)L_2$ as the starting molecule for Ni(I)L (Figure 13).

$$Fe^{III}L_3 + ROOH \rightarrow Fe^{II}L_2 + LH + RO_2^{\cdot}$$

As long as $Ni(II)L_2$ is present in the system, the main decomposition of the hydroperoxide runs by nickel catalytic mechanism [keeping the free radical concentration very low, but after transformation of all $Ni(DTC)_2$ to $Ni(OH)_2$] the peroxide decomposition begins, catalyzed by $Fe(DTC)_3$ with parallel radical release and prooxidant

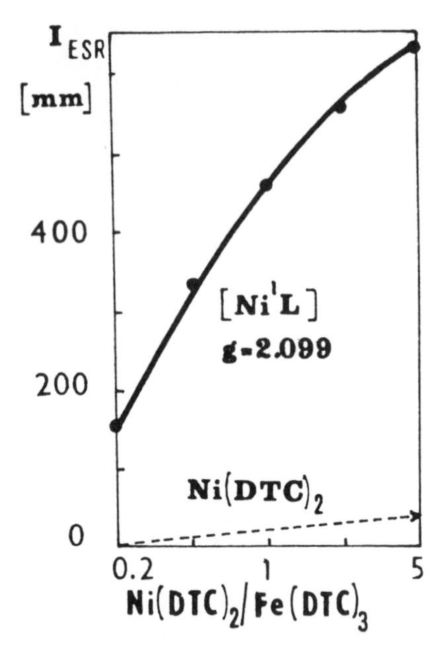

Figure 12. Increase of the intensity of Ni(I) ESR signal with the increase of the molar ratio $Ni(DTC)_2$: $Fe(DTC)_3$. Mixing of 0.25 M benzene solution after adjustation to 0.5 cm^3. Increase of Ni(I) signal with $Ni(DTC)_2$ increase without $Fe(DTC)_3$: dotted line.

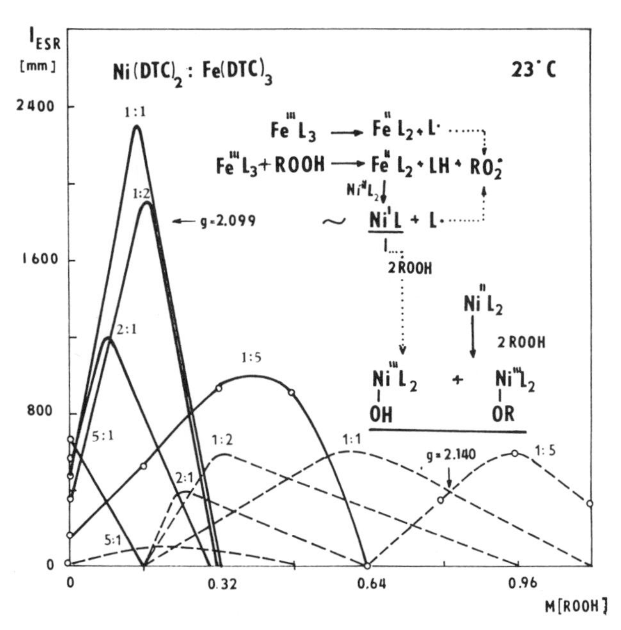

Figure 14. Dependence of ESR signal height of Ni(I)L (full lines) and of $HONi(III)L_2$ or $RONi(III)L_2$ (dotted lines) upon concentration increase of ROOH at different molar ratios of $Ni(DTC)_2$: $Fe(DTC)_3$ in 0.5 cm^3 benzene solution prepared by mixing of 0.25 M solutions.

effect. The peroxide decomposition controlled by Ni(I)–Ni(II)–Ni(III) ET cycle is a sensitive function of the initial concentration of the two components $Ni(DTC)_2$ and $Fe(DTC)_3$ as well as of the actual concentration of ROOH (Figure 14). With peroxide increase in the binary system of $Ni(DTC)_2$ and $Fe(DTC)_3$, the ESR signal of Ni(I)L (g = 2.099) achieves its maximum intensity and is gradually replaced by the signal of Ni(III) (g = 2.14). The maxima of both signals Ni(I) and Ni(III) are dependent on the molar ratio of $Ni(DTC)_2$ and $Fe(DTC)_3$. An intense ESR signal of Ni(I)L indicates an efficiently operating system of peroxide decomposition without es-

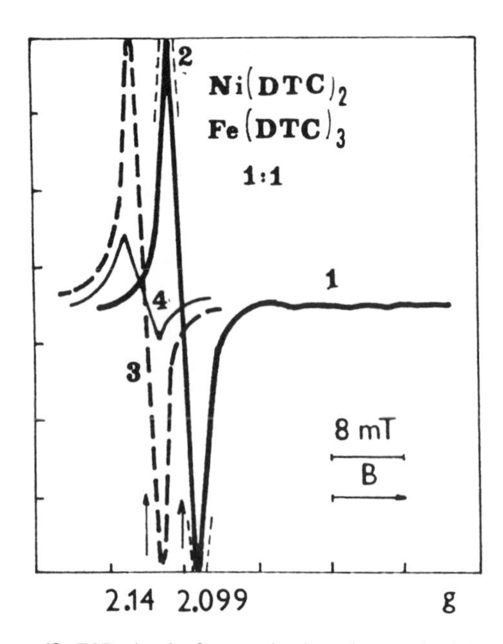

Figure 13. ESR signal of an equimolar mixture of $Ni(DTC)_2$ and $Fe(DTC)_3$ in 0.05 M benzene solution (1), after oxidation with 0.16 M (2) 0.32 M (3) and 0.48 M (4) ROOH.

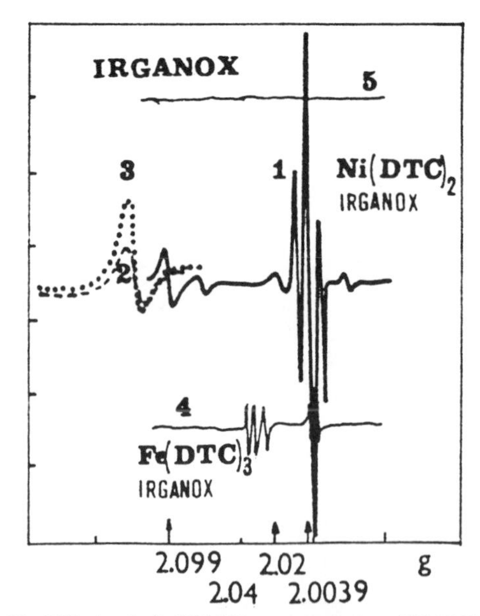

Figure 15. ESR signal of a 0.05 M benzene solution of $Ni(DTC)_2$ in the presence of equimolar amount of Irganox 330 (1), after addition of 0.16 M (2), 0.32 M (3) of ROOH. Equimolar mixture of $Fe(DTC)_3$ with Irganox (4). The solution of Irganox 0.05 M was without signal (5).

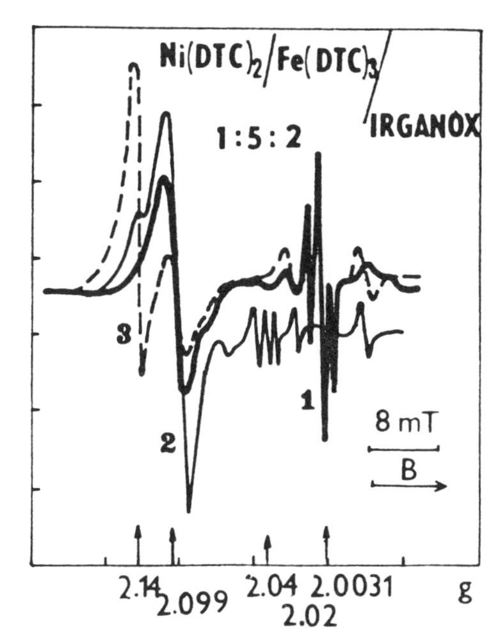

Figure 16. ESR signal of the ternary mixture $Ni(DTC)_2$: $Fe(DTC)_3$: Irganox = 1:5:2. By mixing of 0.25 M benzene solutions of all components, 0.5 cm^3 sample for ESR cell was prepared (1). Change of the signal after adding of 0.16 M (2), 0.32 M (3) ROOH.

cape of reactive radicals, which could initiate oxidative chain reactions.

The efficiency to keep free radicals under control is increased by addition of antioxidants as HT mediators into the binary system. On adding the antioxidant Irganox 330 (0.05 M) to the benzene solution of $Ni(DTC)_2$, the ESR

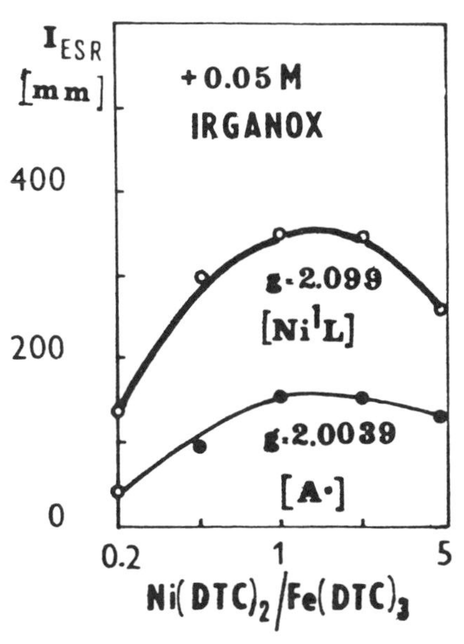

Figure 17. Dependence of the ESR signal heights of Ni(I)L and of phenoxy radicals A· generated from Irganox (0.05 M) upon the molar ratio of $Ni(DTC)_2$:$Fe(DTC)_3$. 0.5 cm^3 of the mixture prepared by mixing of 0.25 M benzene solutions of the individual components.

signal of the phenoxy radical generated from the antioxidant (g = 2.0039) increases effectively, and simultaneously the signal of the ligand radical L· (g = 2.02) disappears in consequence of an HT-LT step (Figure 15).

$$Ni^{II}L_2 \rightarrow Ni^{I}L + L\cdot$$

$$L\cdot + AH \rightarrow LH + A\cdot$$

The supposed internal charge transfer ligand signal of $Fe(DTC)_3$ (g = 2.04) does not abstract H-atom from Irganox.

In the ternary system of $Ni(DTC)_2$, $Fe(DTC)_3$ and Irganox, the Ni(II)-Ni(I) lower redox cycle is better protected against the oxidation with ROOH to the higher Ni(II)-Ni(III) redox cycle and a higher efficiency of radical scavenging is evident with a significant synergetic inhibition effect (Figure 16).

Keeping the concentration of Irganox constant (0.05 M), the highest concentration of Ni(I)L and of phenoxy radicals A· generated from the antioxidants at the equimolar ratio $Ni(DTC)_2$:$Fe(DTC)_3$ = 1 takes place (Figure 17).

$$Fe^{III}L_3 \rightarrow Fe^{II}L_2 + L\cdot \xrightarrow{AH} LH + A\cdot \; (g = 2.0039)$$

$$\downarrow Ni^{II}L_2$$

$$(g = 2.099) \qquad Ni^{I}L_2$$

The post prooxidant effect of catalytic decomposition of peroxides without simultaneous scavenging of free radicals can start only after exhausting the original $Ni(DTC)_2$, which is the key to the programmed inhibition and subsequent initiation of chain oxidation.

EXPERIMENTAL

ESR spectrometer Bruker SRC-200 and Varian E-3 with 100 kHz modulation operating in X-band were used.

Coordinated tert.butyl peroxy radicals were obtained by the reaction under vacuum at 100°C of dried 2% $Co(acac)_2$ [acetylacetonate dicobalt; bis(2,4-pentadionato)Co(II)] dissolved in water-free benzene with tenfold molar excess of tert.butyl hydroperoxide (ROOH, 92% Fluka), at 20–23°C temperature [11–12]. The ESR spectrometer was calibrated according to Varian ESR pitch standard. The ESR cell was filled with 0.3 ml solution of $Co(III)RO_2^{\cdot}$ radicals representing $7 \times 10^{14} \pm 20\%$ spins. 0.3 ml of dissolved antioxidants in molar range of 0.5×10^{-1} to 1.0×10^{-4} M was added under nitrogen screen and, before ESR measurement, stirred for one minute by bubbling with N_2.

When initiation of HT was carried out under exclusion of oxygen, the non-reacted ROOH was evaporated under vacuum at 10°C and then the green powder-like residue was dissolved in benzene. This operation was repeated twice, giving, after the second dilution with 0.3 ml benzene ca 6 × 10^{14} spins of coordinated peroxy radicals. 0.25 M solutions in benzene were used to study the multicomponent mixtures of nickel and iron dipropyl dithiocarbamates or of Irganox 330. The different molar ratios were prepared in ESR cells, where the measured volume was adjusted to 0.5 cm^3 with benzene.

For precise release of bromine or hydrogen bromine into the gaseous phase of pyrolyzed polymers in vacuum, it is convenient to combine aliphatic and aromatic bromine derivatives such as the commercial flame retardant Sandoflam: 4,4-bis(brommethyl)-2,6 dioxa-1-(2,4,6 tribromphenoxy)-1-phosphacyclohexane-1-oxide.

REFERENCES

1. Kevan, L. "Free Radical Study by ESR," in *Methods in Free Radical Chemistry, Vol. I*. E. S. Huyser, ed. New York: Marcel Dekker, p. 27 (1969).
2. Tkáč, A. "A Study of Flame Retardants Mechanism by ESR," in *Developments in Polymer Stabilization, Vol. V*. G. Scott, ed. Essex: Applied Science Publishers LTD, p. 153 (1982).
3. Tkáč, A. and V. Kellö. *Rub. Chem. Technol.*, 28:383 (1955).
4. Tkáč, A. and V. Kellö. *Rub. Chem. Technol.*, 30:1255 (1957).
5. Semjonov, N. V. *Usp. Chim.*, 20:673 (1951).
6. Tkáč, A. *Ind. J. Radiat. Phys. Chem.*, 7:457 (1975).
7. Tkáč, A. "Coordinated Radical on Hemoproteins in the Course of Catalytic Decomposition of Hydroperoxides," in *Fundamental Research in Homogenous Catalysis, Proceedings of the IV International Symposium on Homogenous Catalysis*. E. A. Shilov, ed. London: Gordon and Breach Science Publishers LTD, p. 817 (1986).
8. Howard, J. A. "Homogenous Liquid Phase Antioxidants," in *Free Radicals, Vol. II*. J. K. Kochi, ed. New York: Wiley-Interscience Publ., p. 54 (1973).
9. Holdsworth, J. D., G. Scott and D. W. Williams. *J. Chem. Soc.*, 4692 (1964).
10. AL-Malaika, S., A. M. Marogi and G. Scott. *J. Appl. Polymer Sci.*, 31:685 (1986).
11. Tkáč, A., K. Veselý and L. Omelka. *J. Phys. Chem.*, 75: 2575 (1971).
12. Tkáč, A. and L. Omelka. *Org. Magn. Reson.*, 14:109 (1980).

B. A. Wolf[1]
M. Ballauff[2]
F. K. Herold[3]

Thermodynamically Induced Shear Degradation of Dissolved Polymers

ABSTRACT

If moderately concentrated polymer solutions are subjected to shear, many polymers suffer degradation even under laminar flow conditions (if the thermodynamic quality of the solvent is sufficiently poor). Chains are however not broken under the same conditions with good solvents. By means of the exact solution of the kinetic equations, the distribution of rupture sites within the polymer molecules was determined from the molecular weight distributions observed at different degradation times. The bonds are found to break highly preferentially in the middle of the chain; the probability of scission is given by a Gauss function with its maximum at the midpoint. From an analysis of the viscoelastic behavior of the solutions, it is concluded that the pronounced reduction in the mobility of the polymer upon the deterioration of the solvent—as well as an increase in the energy the system can store—causes the thermodynamically induced shear degradation. Furthermore the knowledge of the viscoelastic behavior of these solutions allows one to predict how sensitive a certain polymer will be to the present type of degradation.

KEY WORDS

Polymer solutions, polymer degradation, rupture site distribution, storable energy, chain entanglements, viscoelasticity, flow curves, poly(n-alkyl methacrylates), poly(styrene).

INTRODUCTION

The present study was initiated quite a few years ago by the following experimental observation [1]. A polymer solution for which the phase separation temperature had been determined was sheared over night by means of a magnetic stirrer. On the next morning it was impossible to reproduce the cloud point temperature measured before. Experiments, subsequently designed to clarify this strange behavior have demonstrated that it is due to a reduction of the molecular weight of the polymer, caused by shear degradation. It was apparent that the polymer chains reacted very sensitively towards shear degradation at the low thermodynamic quality of the solvent. The following facts were ascertained from these early experiments: when a polymer solution is sheared at a temperature only a few °C distant from its demixing temperature, even moderate shear rates, which guarantee *laminar* flow, lead to chain scission. In contrast, when the same solution is sheared at higher temperatures under thermodynamically more favorable conditions, or when a solution of identical polymer concentration in a good solvent is sheared at any temperature, even drastically larger shear rates cannot break the chains.

Since this early, more preliminary study, the question was whether the phenomenon of the thermodynamically induced shear degradation is a general one (i.e., observable with all polymers). How it can be rationalized in molecular terms has been of particular and constant interest to us. In this paper it is reported to what extent these questions can at present be answered. Since part of this material has already been published [2–4], only the new aspects are presented in full detail, and the rest of the information is restricted to the minimum necessary for understanding the main ideas. Furthermore, the present article excludes non-thermodynamically induced degradation phenomena—reviews on this subject can be found in the literature [5].

EXPERIMENTAL

Materials

The *polymers* under investigation were poly(styrene) (PS) the number average degrees of polymerization P_n ranging from 2,650 to 15,000 and $U = (P_w/P_n) - 1$ from 0.06 to 0.7, poly(methyl methacrylate) (***PMMA***), $P_n = 6{,}900$ and $U = 0.20$ poly(n-butyl methacrylate) (***PBMA***) $P_n = 14{,}500$ and $U = 0.23$, and, finally, poly (n-decyl methacrylate) (***PDMA***) $P_n = 3{,}200$, $U = 0.29$.

[1]Institut für Physikalische Chemie der Universität, D-65 Mainz, FRG.
[2]Max-Planck-Institut für Polymerforschung, D-65 Mainz, FRG.
[3]Hoechst AG, D-6230 Frankfurt, FRG.

Solvents of the highest available grades were employed. The following served as theta solvents: *trans*-decalin (***TD***) for PS, heptanone-3 (***HEP-3***) for PMMA, 2-propanol (***2-POH***) for PBMA and 1-pentanol (***1-POH***) for PDMA. As thermodynamically good solvents the following were used: toluene (***TL***) for PS and PMMA, methyl ethyl ketone (***MEK***) for PBMA and *iso*-octane (***IO***) for PDMA.

Since free radicals are formed in the process of shear degradation, *radical scavengers* had to be added to the polymer solutions in order to prevent a recombination of the fragments. Depending on their solubility in the solvents of interest, DPPH or Galvinoxyl were used. Some of the experiments were performed in the presence of oxygen since it turned out to suffice as a radical scavenger.

Apparatus

In order to study the *rheology* of the solutions of interest, several rotational viscometers of the Searle and of the cone-and-plate type (Haake Meßtechnik, Karlsruhe) were used.

The *shear cell* in which the degradation experiments were performed consists of a Couette-type apparatus with a bell-shaped rotor and a good thermally insulated stator. It can be operated in the presence of a protective gas, like nitrogen, and reaches maximum shear rates on the order of 11,000 s^{-1}. Samples are taken from the outer gap by means of a syringe inserted through a septum.

The molecular weight distribution (m.w.d) of the polymers was determined by means of *gel permeation chromatography* (high pressure GPC, Waters). The access to exact data is an essential prerequisite for studying the detailed degradation kinetics and in particular the distribution of rupture sites within the individual polymer molecules.

RESULTS AND DISCUSSION

Flow Curves

In order to obtain information on the viscosities of the polymer solutions under the conditions of the degradation, measurements were performed for all the systems under investigation. Figure 1 gives an example for HEP-3/PMMA.

The solid line in Figure 1 represents a theoretical curve, adjusted to the experimental data according to a concept recently developed for the description of flow curves of polymer solutions under thermodynamically very unfavorable conditions [6]. In this concept, the measured viscosity is split into an entanglement part η_{ent}, quantitatively described by the Graessley entanglement theory, and η_{fric}, a contribution which is independent of

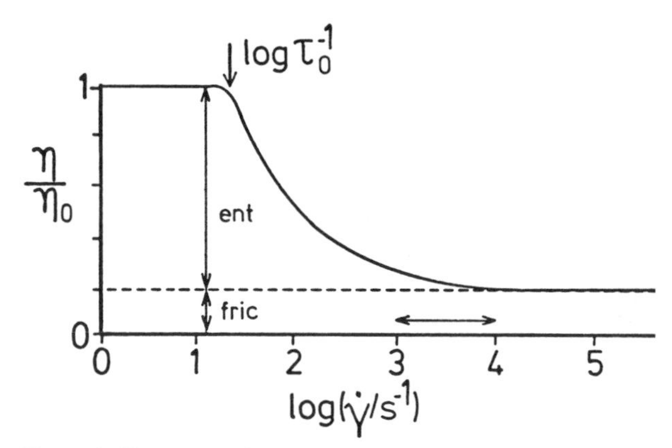

Figure 1. Flow curve of a 6.81 wt. % solution of PMMA in HEP-3 at 25°C, i.e., 1°C above the phase separation temperature of this solution. Note that this graph is on a semi-logarithmic scale, instead of the normal double-logarithmic one, so that η_{ent} and η_{fric} can easily be split (cf. text). The shear rate corresponding to the τ_0 value obtained from the theoretical evaluation of η_{ent} is indicated. The horizontal arrow shows where the degradation experiments are normally performed.

shear rate. From the variation of η_{ent} with shear rate, τ_0, the characteristic relaxation time associated with the onset of shear rate dependence in the visocosity can be obtained in the normal way. η_{fric}, the non-entanglement frictional contribution, on the other hand, accounts for the fact that as the solvent quality becomes very poor, the number of contacts between the polymer segments increases drastically. This leads to an elimination of the ball-bearing-like or lubricating effect of the solvent. As a result, an additional mode of energy dissipation leading to a contribution to the viscosity, independent of shear rate, becomes active.

The analysis of τ_0 and η_{fric} data with respect to thermodynamic influences reveals that the preference of intersegmental contacts in poor solvents normally leads to a very pronounced ascent of both quantities.

Figure 2 shows how the relaxation time, characteristic for entanglement processes, increases drastically in the case of the system TD/PS, as the phase separation temperature T_D is approached; for comparison this dependence is also given for a solution of PS of equal concentration in the good solvent TL. In this case τ_0 rises only slightly upon a reduction of temperature.

Among the polymers under investigation PDMA represents an exception. Solutions of this polymer always behave as if the thermodynamic interactions were very favorable, even in theta solvents close to phase separation. A hypothesis for this behavior will be given in the discussion.

Degradation

OVERALL EFFECTS

The majority of experiments was performed at moderate polymer concentrations of typically five-fold chain overlap ($c_2 \cdot [\eta]$ ca. 5, c_2 being the polymer concentra-

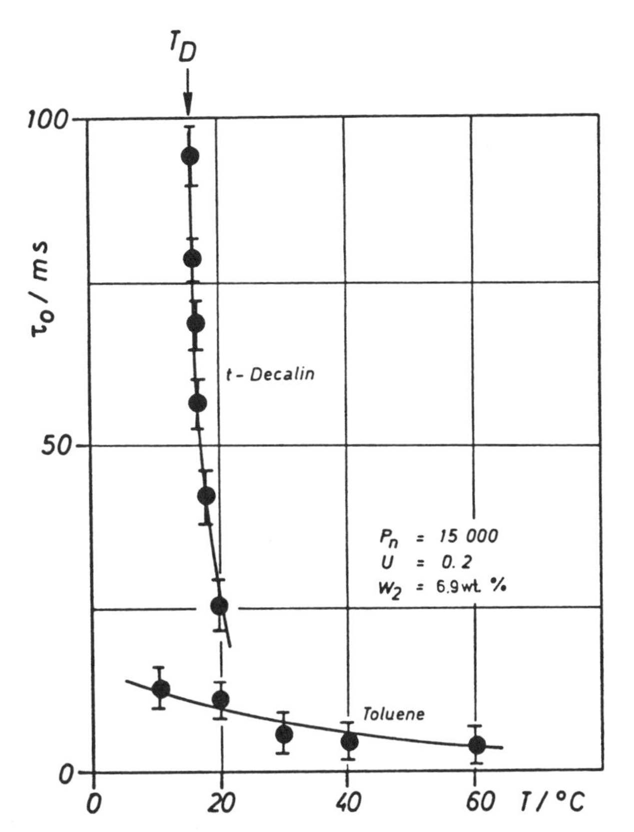

Figure 2. Variation of τ_0 with temperature for a 6.9 wt.% solution of PS in the theta solvent TD and in the good solvent TL.

tion in g/cm³ and $[\eta]$ the Staudinger-Index in cm³/g), up to shear rates of ca. 20,000 s^{-1}, at a temperature distance ΔT from the particular demixing temperature (cf. Figure 3) ranging from 0.2 to 20 K.

The polymer degradation reached after a certain time t of shear was measured in terms of B_t, the average number of bonds broken per molecule within the time interval t. B_t can be calculated from the measured degrees of polymerization according to

$$B_t = (P_{n,0} - P_{n,t})/P_{n,t} \quad (1)$$

The increase of B_t with time can be well represented by the following equation [7,8]

$$(dB_t/dt) = k(B_\infty - B_t) \quad (2)$$

in which B_∞ signifies a limiting number of broken bonds, corresponding to infinitely long degradation times. This quantity can easily be determined from $\dot{B}_0$, the initial degradation rate, since B_t is zero at the beginning of the experiment.

In the absence of radical scavengers no degradation is observed with any of the polymers. In the presence of such additives, all polymers under investigation turned out to be sensitive to the thermodynamically induced shear degradation except PDMA, which did not degrade in the vicinity of phase separation, not even at the highest accessible shear rates and polymer concentrations. For the polymers which do show chain scission, i.e., PS, PMMA, and PBMA, the observed B_∞ values ranged from 0.1 to 1 per molecule, and the $\dot{B}_0$ values from 0.01 to ca. 2 per molecule per hour depending on the particular system, polymer concentration, distance from the demixing temperature, and shear rate.

The initial degradation rate increases approximately with the third power of polymer concentration and with the fifth power of shear rate. One unexpected finding seems noteworthy—if $\dot{B}_0$ is plotted as a function of T for a given polymer solution at constant shear rate, it does not increase monotonically upon approaching the two phase region, but exhibits a maximum close to the demixing condition.

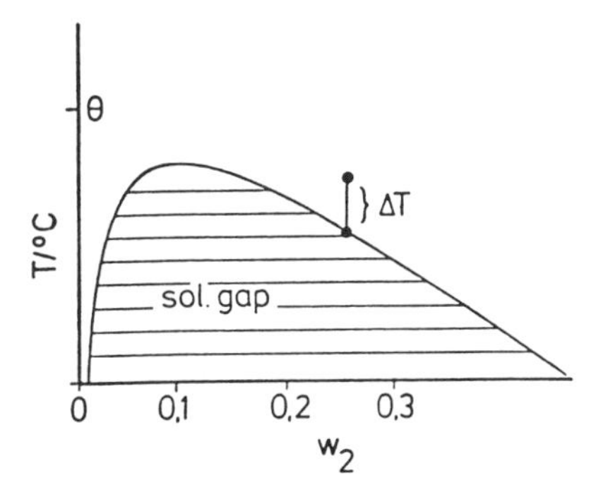

Figure 3. Sketch demonstrating where the degradation experiments in thermodynamically poor solvents are typically performed. Θ is the theta temperature and ΔT gives the distance of the degradation temperature from the two phase region.

Degradation

ANALYSIS OF THE ***m.w.d.***

In addition to the overall effects discussed so far, the information obtained from GPC measurements was used to study at which positions of the individual chains the chemical bonds are typically broken and to what extent the longer chains present in a polymolecular sample are broken preferentially. By means of GPC measurements it was also demonstrated that chain scission actually does take place in the absence of radical scavengers. In this case no change in P_n can be observed ($B = 0$) since the polymer fragments recombine, however, the m.w.d. does normally change—with sufficiently narrow distributions of the starting material it broadens upon degradation.

Figure 4 contains the m.w.d. of an original polymer (PS of low molecular non-uniformity, $P_{n,0} = 2,600$) and of that product after six hours of shearing, resulting in a rupture of 30% of all molecules ($B = 0.3$). From the oc-

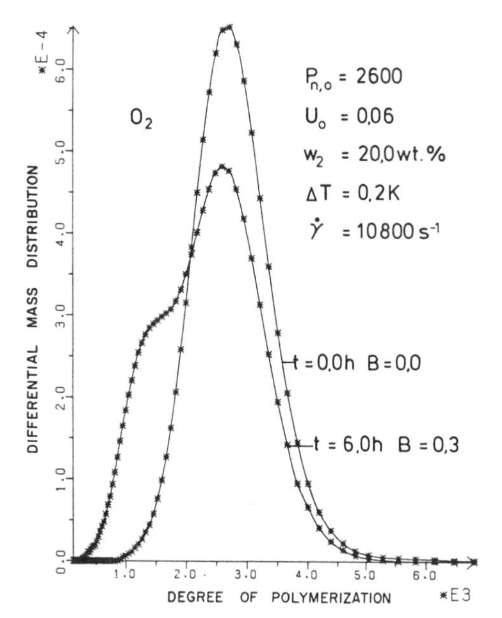

Figure 4. Example of a differential m.w.d. of a starting polymer and of a degraded product, given for the system trans-decalin/poly(styrene). After shearing a solution of 20 wt. % of the polymer 0.2 K above its demixing temperature at a shear rate of 10,800 s^{-1} for 6 hours, 3 molecules out of 10 have on the average suffered chain scission.

currence of a second maximum in the m.w.d. of the degraded sample at approximately half the initial P_n it becomes immediately obvious that the chains tend to break preferentially in the middle. On closer inspection one can also see that the higher molecular weight side of the original distribution is affected somewhat more by the degradation than the lower.

A more detailed evaluation of the m.w.d.s at different degradation times by means of an exact solution of the kinetic equations [2] quantifies to the above observations: Let i be the number of subunits into which a chain could theoretically be broken, then k_{ij}, the probability that this chain ruptures into two pieces containing j and $(i - j)$ subunits, respectively, can be written as

$$k_{ij} = k(i - 1)^x (Ri)^{-1} (2\pi)^{-0.5} \exp[-0.5(j - i/2)^2 (Ri)^{-2}] \tag{3}$$

The term $k(i - 1)^x$ gives the total degradation rate constant of the molecule with i subunits, and the remainder of the r.h.s. of the above equation [9] constitutes a Gauss function, with its maximum at the midpoint of the chain. R, the characteristic parameter in this rupture site distribution (r.s.d.), is the normalized (divided by i) standard deviation.

By means of the mathematical procedure mentioned above it is possible (for a given m.w.d. of the starting polymer) to fit the two model parameters, x and R, to the experimental data such that the changes in the m.w.d.s with increasing degradation time can be reproduced quantitatively using only these two constants.

The x values obtained by these fits range from 0.1 to 0.7 and are therefore considerably lower than predicted [9]; according to the present data x decreases with increasing polymer concentration. R is found to vary from 0.03 to 0.20. It depends on the degradation conditions in a very complex manner. The r.s.d. is shown in Figure 5 for some typical R values.

From the above graph it can be seen that the chain cleavage takes place almost exclusively in the middle of the molecule in those experiments characterized by $R = 0.03$. With $R = 0.15$ the probability that the chain breaks into two equal halves is still more than twice as large as for a rupture of 1/3 of the molecule. With $R = 1.5$, however, the degradation has already become practically indistinguishable from statistical.

Molecular Models

For the present type of degradation to occur, stresses have to be built up during shear that are large enough to result in a cleavage of C–C bonds. Chain entanglements are commonly assumed responsible for the development of these forces. The fact that polymers dissolved in good solvents remain stable under conditions for which pronounced shear degradation is observed in poor solvents, indicates that not any type of entanglement suffices but only those which are sufficiently long-lived. As already briefly mentioned in the context of flow curves, the reduction in molecular mobility in poor solvents is due to a thermodynamic preference of intersegmental contacts over contacts between solvent molecules and polymer segments. This situation leads to a sort of "dry" friction between the different polymer strands and increases the average life-time of entanglement points, measured by τ_0,

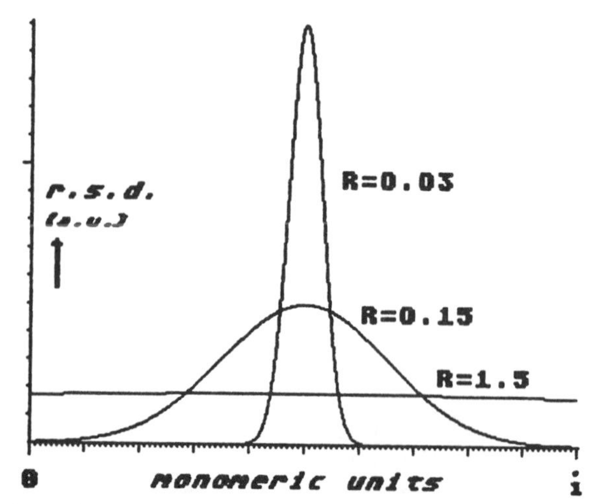

Figure 5. Typical rupture site distributions calculated according to Equation (3) with the indicated normalized standard deviations R.

to such an extent that they eventually become grip-points which can be used to split the chains.

For a better understanding of the degradation mechanism, the energy required to split the polymer backbone as compared to the energy introduced into the system by shearing must be known. Corresponding calculations demonstrate that the energy required to split the chains normally corresponds to less than 10^{-6} of the dissipated energy. This fact and the intuition that the presumably undirected stresses resulting from energy dissipation would not lead to chain rupture, raises the question whether the energy stored during stationary flow could be the decisive quantity, as it is in the thermodynamics of flowing polymer solutions [10]. In the following section some model calculations are performed in order to check this idea.

E_s, the energy a system is able to store per volume during stationary flow can be calculated [11] for linearly viscoelastic systems as

$$E_s^\circ = G_0^{-1} \eta_0^2 \dot{\gamma}^2 \tag{4}$$

where G_0 is the shear modulus of the solution, η_0 the zero shear viscosity and $\dot{\gamma}$ the shear rate. By means of the Maxwell equation

$$G_0 = \eta_0 / \tau_0 \tag{5}$$

Equation (4) transforms to

$$E_s^\circ = \tau_0 \eta_0 \dot{\gamma}^2 \tag{6}$$

The above relation makes it at once obvious that the ability of polymers to store energy increases as the relaxation times become longer, i.e., as the solvent quality goes down.

Selecting the variables of Equation (6) to be typical for the present experiments, one obtains E_s values on the order of 10^4 to 10^5 J/m^3, depending on the particular manner in which the non-linear viscoelastic behavior is accounted for. By multiplying this value by the volume of an individual polymer coil, the energy stored inside the domain of a macromolecule can be calculated. The product of $[\eta]_\Theta$ (the Staudinger-Index for theta conditions) and the molecular weight of the polymer yields a good estimate of the molar volume of coils in the moderately concentrated solutions of present interest. Dividing this value by Avogadro's number leads typically to 10^{-22} m^3 per molecule. From the above figures one can calculate that 10^{-18} to 10^{-17} J of mechanical energy are stored within a single polymer molecule. The question now arises whether this quantity suffices, if properly focused, to break a C–C bond. Since it takes 0.6×10^{-18} J to break such a junction, the answer is yes.

Still open however are the mechanisms of concentrating the stored energy into just a few sites. Within the framework of the present entanglement model, it is reasonable to assume that the grip-points mentioned above constitute the loci where the chains are broken. For this assumption to be true the number of entanglements would have to be rather small, since the energy could otherwise be distributed on too many bonds and would therefore not lead to chain scission. Model calculations concerning the number of entanglements start from published entanglement molecular weights for polymer melts. Dividing the polymer molecular weight by this value one obtains the equilibrium number of entanglements per polymer molecule. Multiplication with the volume fraction of the polymer contained in the solution yields the equilibrium number of entanglements of dissolved molecules, from which the lower values at the shear rates where the degradation takes place can be calculated by means of the g-function of Graessley's theory [12].

For the polymers and conditions of interest, the calculations along the above lines lead to typically 1–2 entanglements per molecule. It can therefore be concluded that it may well be the energy stored in flowing polymer solutions which, when focused to only a few bonds, leads to the thermodynamically induced shear degradation.

CONCLUSIONS

The investigations presented have shown how the thermodynamic situation in poor solvents normally leads to a dry friction of polymer strands, which reduces the molecular mobility and increases the energy the system can store while flowing. By a proper concentration of this energy into just a few grip points within a polymer molecule, the chains break. The analysis of the m.w.d. of the degraded products reveals a high preference of central cleavages.

Although the thermodynamically induced shear degradation appears to be a very general phenomenon, the measurements with PDMA have demonstrated that exceptions are possible. Poly(n-decyl methacrylate) does not degrade, even under the most unfavorable thermodynamic conditions. This finding is in accord with the present molecular model, since the mobility of this chain (as expressed by the viscometric relaxation time τ_0) turns out to be unaffected by the quality of the solvent—which means that the system cannot store sufficient energy. The obvious reason for this behavior lies in the highly flexible decyl side groups pending on the polymer backbone and acting like constantly attached solvent molecules. The dry friction causing degradation is consequently made impossible.

According to the present results the sensitivity of a certain polymer against thermodynamically induced shear degradation can be predicted from the viscoelastic properties of its solutions in thermodynamically poor solvents.

ACKNOWLEDGEMENTS

We are grateful to the AIF (Arbeitsgemeinschaft industrieller Forschungsvereingungen eV) for its financial support.

REFERENCES

1. Breitenbach, J. W., B. A. Wolf and J. K. Rigler. *Makromol. Chem.*, 164:353 (1973).
2. Ballauff, M. and B. A. Wolf. *Macromolecules*, 14:654 (1981).
3. Ballauff, M. and B. A. Wolf. *Macromolecules*, 17:209 (1984).
4. Herold, F. K., G. V. Schulz and B. A. Wolf. *Polymer Commun.*, 27:59 (1986).
5. Casale, A. and K. S. Porter. *Polymer Stress Reactions*. NY:Academic Press (1978). W. Schnabel. *Polymer Degradation*. München:Hanser Intern. (1981).
6. Ballauff, M., H. Krämer and B. A. Wolf. *J. Pol. Sci., Phys. Ed.*, 21:1217 (1983).
7. Jellinek, H. H. G. and G. White. *J. Polym. Sci.*, 7:33 (1951).
8. Glynn, P. A. R., B. M. E. van der Hoff and P. M. Reilly. *J. Macromol. Sci., Chem.*, A6:1653 (1973).
9. Bueche, F. *J. Appl. Polym. Sci.*, 4:101 (1960).
10. Wolf, B. A. *Macromolecules*, 17:615 (1984).
11. Vinogradov, G. V. and A. Ya. Malkin. *Rheology of Polymers*. Moscow:Mir Publishers. Berlin, Heidelberg, New York:Springer Verlag (1980).
12. Graessley, W. W. *Adv. Polym. Sci.*, 16:1 (1974).

H. ZIMMERMANN[1]

Degradation and Stabilization of Poly(Alkylene Terephthalates)

ABSTRACT

The thermal and thermooxidative properties of polyesters like poly(ethylene terephthalate) are essentially affected by temperature, metal derivatives used as transesterification catalysts and by modification of the chemical structure caused by side reactions or used to alter specific properties of the polymer.

The kinetics of thermal degradation of PET and PBT were investigated by determining the rate of increase of carboxyl groups. On the basis of these results, polymer analytical investigations, and model reactions, the mechanism of thermal degradation and catalysis by metal derivatives as well as possible methods of improving thermal stability, is discussed.

Thermooxidative degradation of PET and copolyesters, as well as the efficiency of antioxidants, has been studied especially by isothermal and nonisothermal thermal analysis (DTA, TGA).

The use of these methods for kinetic studies is illustrated.

INTRODUCTION

The ever-increasing use of polymers and the demand for extended lifetime make it necessary to understand in detail the specific mechanism of polymer degradation and the factors influencing it, in order to find suitable ways to minimise such reactions. Seldom does a polymer exist in a completely pure state to make it possible or reasonable to investigate the pure polymer. To come up to the demands of synthesis or to the specific demands connected with the different applications it is necessary to use special additives, or to overcome certain shortcomings, one must modify the macromolecule. Additives present in polymers are, e.g., catalysts, stabilizers, emulsifiers, plasticisers, nucleating agents, fillers, and reinforcement agents. All these additives and co-components, as well as small quantities of impurities or irregularities in the polymer chain formed during synthesis can interact and markedly change the behavior of the polymer. Such problems of polymer stability can be demonstrated by the example of poly(alkylene terephthalates), especially poly(ethylene terephthalate) (PET).

Some examples of PET modifications by formation of copolyesters to get special properties are shown in the following table:

Product	Aim	Modifying Agent
fibers	dyeability, high-shrinkage, low-pilling, elasticity	polyalkylene glycol, sulfo-isophthalic acid adipic-, sebacic-, isophthalic acid
thermoplastic elastomers	rubber-like elasticity	polyethylene glycol, polybutylene glycol
melt adhesives	elasticity; $T_m\downarrow$; $T_g\downarrow$	butanediol, hexanediol, adipic acid
amorphous, transparent PET	suppression of crystallization	cyclohexane-1-4-dimethanol, isophthalic acid
injection moulding PET	increase of crystallization properties	3-methyl pentanediol, 2-methyl hexanediol, butanediol

EXPERIMENTAL

The polyester samples investigated were prepared in a laboratory device by polycondensation of bis-(2-hydroxyethyl)terephthalate (BHT) in the presence of the appropriate metal compounds, stabilizers, or co-components. For comparison, commercial samples were investigated too.

The carboxyl group concentrations were determined for samples dissolved in benzylalcohol using a modified version [1] of the method of Pohl [2].

The solution viscosity η_{rel} was determined in a mixture of phenol/tetrachlorethane at 20°C using an Ubbelohde viscosimeter.

[1]Institute of Polymer Chemistry "Erich Correns," Academy of Sciences of the GDR, Teltow-Seehof, German Democratic Republic.

The concentrations of metal ions and phosphorus compounds present in the polyester were determined by several chemical microanalytical methods [3,4].

The content of diethylene glycol (DEG), incorporated into the polyester chain, was determined by gas chromatography after hydrazinolysis of PET [5].

Degradation experiments were made directly in the polycondensation apparatus (open system) or under nitrogen pressure (closed system) [6].

Differential Thermal Analysis (DTA) and Thermogravimetric Analysis (TGA) were carried out using a Mettler Thermoanalyzer with a sample weight of 20 mg, a flow rate of air or N_2 of 7 litres h^{-1} and several heating rates. The polyester samples were cooled in liquid nitrogen, ground, screened and dried. A particle size fraction of 0.12–0.25 mm was chosen.

Investigations on photochemical degradation by irradiation of PET films were performed using a Xenotest 1200 accelerated weathering tester (4.5 kW Xenon burner; wavelength $\lambda > 280$ nm) and the increase in carboxyl group content was measured. The temperature in the chamber was about 30°C. The humidity was not determined exactly, however there was no difference between results of pure irradiation and weathering by applying a 15 min rain/15 min dry cycle.

Thermal Degradation

The main reactions of degradation and factors influencing the thermal stability of PET are rather well known [1,7–12]. Some of them which have to be regarded all the more in order to meet demands arising from new kinds of preparation, processing, or application of PET are discussed in the following presentation.

In the preparation of PET the ester interchange of dimethyl terephthalate (DMT) and the melt polycondensation of BHT in vacuum at 270–290°C must be catalytically accelerated. Effective catalyst derivatives of Ca, Mg, Pb, Co, Zn, Mn, Sb, Ti and Ge are used in the range 1×10^{-4}–5×10^{-4} moles per mole DMT or terephthalic acid. These metal compounds affect not only the rate of polycondensation, they also promote certain undesirable side reactions, in particular the formation of ether links incorporated into the polymer chain and the thermal degradation of the polyester formed.

Let us discuss the main steps of the complex degradation process of PET. The initial step is a random scission of the chain at an ester linkage, resulting in the formation of a vinyl ester end-group and a carboxyl end-group [1]. We could prove that this reaction step is accelerated to various extent by the metal compounds [1,10]. This reac-

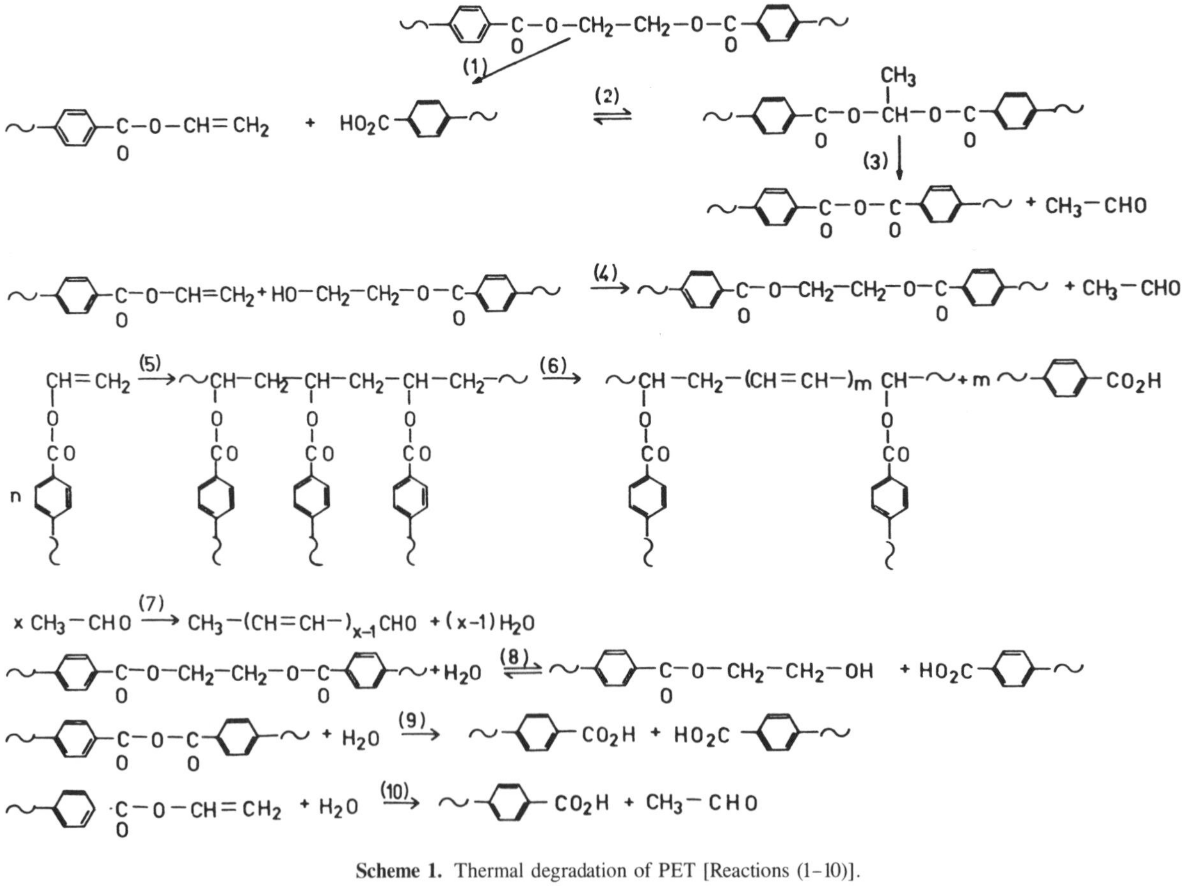

Scheme 1. Thermal degradation of PET [Reactions (1–10)].

tion proceeds permanently also under the conditions of polycondensation. As long as the end-groups of PET are predominantly hydroxyl ester groups, the vinyl ester end-groups undergo transesterification [Reaction (4)].

In this way polyester chains are regenerated and the average degree of polymerisation is maintained. The only consequence of Reactions (1) and (4) is the replacement of a hydroxyl end-group by a carboyxl end-group and the production of an equivalent amount of acetaldehyde. The stability or the decrease of the average molecular weight depends mainly on the rate of these two reactions.

When the number of hydroxyl end-groups has decreased to a large extent either by further polycondensation or by (4), Reactions (2) (3) and (5) (6) can increase as parts of the overall process. The progressive discoloration of PET during degradation can be attributed to Reactions (5) (6).

In addition to these main reactions of thermal degradation various other reactions may occur. Many different minor degradation products are formed. Among the gaseous products acetaldehyde amounts to 80% of the total formed especially by Reactions (4) and (2) (3) [10,13].

The manufacture of PET containers has developed worldwide into a high volume [14]. Retained acetaldehyde in PET packaging materials can migrate to the foods and cause changes in taste [15,16]. Depending on the degree of thermal degradation during production and manufacture of PET the residual content of acetaldehyde may amount to about 100 ppm. By a solid state post-polycondensation at 200–240°C, performed mainly to increase the molecular weight of PET up to a level needed for technical use, the content of acetaldehyde can be—and must be—reduced to a value <5 ppm. Here you can see a special need to prevent degradation as far as possible, arising from a new application of PET.

The literature data on the kinetics of thermal degradation of PET depend on the method of evaluation (decrease of melt or solution viscosity, increase of carboxyl end-groups, weight loss by volatile degradation products) [1,6,9,11,17]. Furthermore, it can be seen from the reaction scheme of the secondary reactions and the transport processes of volatile reaction products, that the resulting experimental data used for the calculation of kinetic data, depend on the reaction conditions. The pretreatment of PET (air-dried or vacuum-dried samples) may also influence the measured thermal stability [18]. We calculated the rate constants of degradation from the increase of COOH-groups by a first-order rate law:

$$\ln \frac{A_0}{A_0 - X} = kt \tag{1}$$

A_0 is the number of ester groups liable to scission, which amounts to 10374–10392 μeq/g PET and $X =$ μeq COOH/g PET measured.

The polymer properties of a PET containing more than 100–150 μeq COOH/g PET are already reduced so far that it is of no technical value, and furthermore—with increasing degradation—side reactions could become more and more important. Therefore, our investigations on degradation kinetics were limited to this initial stage, up to about 3% conversion of the ester groups.

Table 1 shows the values of the rate constants and activation parameters on the increase of COOH of various samples. The accelerating effect of the Zn compound used as catalyst is obvious. The much higher negative value of $\Delta S^{\neq}$ in the presence of a Zn-catalyst is noteworthy.

With reaction conditions I the reaction products, ethylene glycol, water and acetaldehyde, were removed by the N_2 stream. Thus, the chain combining reactions compete with the chain scission, and the data are composite constants for the complex processes.

Table 1. Rate constants and activation parameters for increase in carboxyl group content on thermal degradation of PET in a stirred (I) and an unstirred (II) open system with flowing N_2 and in a closed system (III).

	PET Sample	Catalyst/Inhibitor 10^{-4} wt.% in PET	$k \cdot 10^6$, s^{-1} 280°C	290°C	300°C	$k \cdot 10^5$, s^{-1} 340°C	E_A kJ mol^{-1}	ln A (s^{-1})	$\Delta S^{\neq}$ J mol^{-1}K^{-1}
I	1	139 Sb	0.15	0.28	0.52	0.47	163	19.72	−94
	2	41 Mn/ 261 Sb/ 42 P	0.12	0.20	0.40	0.41	170	20.95	−85
	3	92 Zn	0.23	0.35	0.57	0.31	122	11.25	−166
II	1	139 Sb	0.27	0.55	1.04	1.39	180	24.13	−58
	4	20 Zn/ 160 Sb	0.42	0.78	1.46	1.40	164	21.0	−84
III	1	139 Sb	0.44	0.92	1.77	1.87	187	25.84	−42
	2	41 Mn/ 216 Sb/ 42 P	0.44	0.87	1.78	1.82	186	15.58	−46
	3	92 Zn	0.65	1.19	2.10	1.81	153	19.1	−100
	1.1	139 Sb/ 60 Zn	0.56	0.99	1.81	1.50	156	19.5	−99
	1.2	139 Sb/ 120 Zn	0.72	1.28	2.10	1.49	141	16.55	−121
	1.3	139 Sb/ 180 Zn	0.91	1.50	2.43	1.51	129	14.21	−141

Table 2. Kinetic data for thermal degradation of PET calculated from nonisothermal TGA measurements by the Zsako/Šatava method (heating rate q:0.5 K/min; weight loss ≤ 1.5%).

PET Sample	Catalyst/Inhibitor 10^{-4} wt.% in PET	E kJ mol^{-1}	ln A (s^{-1})	$k \cdot 10^6$, s^{-1}			$k \cdot 10^5$, s^{-1}
				280 °C	290 °C	300 °C	340 °C
1	139 Sb	304	50.33	0.14	0.45	1.41	0.90
2	41 Mn/ 216 Sb/ 42 P	325	54.51	0.10	0.33	1.12	0.96
3	92 Zn	271	43.95	0.31	0.88	2.43	0.99
4	20 Zn/ 160 Sb	298	49.16	0.16	0.51	1.55	0.91
5	100 Mn/ 248 Sb/ 10 P	291	47.87	0.20	0.62	1.83	0.92
6	70 Mn/ 76 Ge/ 79 P	282	46.07	0.24	0.71	2.02	0.96

In the preparation and processing of PET (especially in continuous processes) the melt usually stands for some time in pipes. In such a closed system as simulated by conditions III the emission of acetaldehyde or water is inhibited. In consequence of this situation, the effective rate of thermal degradation in the closed system is significantly higher than that in the open system, but the influence of the catalyst on degradation is the same.

We have also studied thermal degradation by isothermal and dynamic thermogravimetric analysis. Table 2 presents the kinetic data. E and A were determined from non-isothermal TGA curves using an evaluation method proposed by Szako [36] and Satava [37]. This method will be discussed later.

The calculated rate constants differ somewhat from the values calculated from the increase in carboxyl groups under isothermal conditions. The order of magnitude of the rate constants and the strong dependence on the catalysts however clearly is the same. An essential aspect should be emphasised. The rate determining cooperation of catalyst and temperature in degradation of PET is most marked in the temperature range necessary for melt polycondensation and processing of this polyester. At temperatures above 300°C, the differences in acceleration by metal compounds progressively decrease.

A probable mechanism of metal ion catalysis can be discussed on the basis of these kinetic data and some model experiments. The same order of acceleration by metal compounds as in the thermal degradation of PET was observed in the thermal degradation of ethylene glycol dibenzoate (EB), which can be considered to be a characteristic segment of the polyester chain. By contrast, in the thermal degradation of diethylene glycol dibenzoate (DB) or 1.4-butylene dibenzoate (BB) no catalysis by metal ions was detected. The activation parameters correspond to those of the uncatalysed thermal scission of other esters [12].

Generally, the pyrolysis of esters is believed to proceed via a cyclic transition state (a). The negative entropy of activation shown in Table 1 can be explained by the formation of such an intermediate cyclic compound. With the Zn-catalysed PET and with EB + $Zn(CH_3COO)_2$, however, the calculated activation entropy was found to be about twice that of antimony-containing PET or EB without catalyst. This result suggests the participation of Zn or other metal compounds, but not Sb, in the activated complex of the transition state, thus favouring the

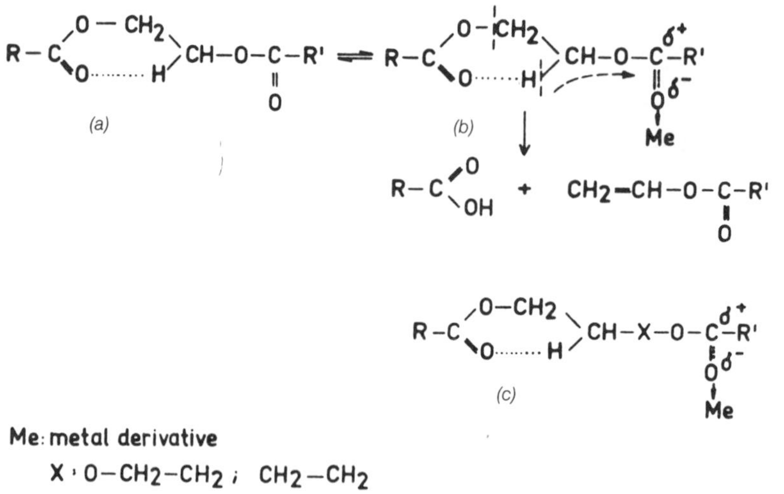

Me: metal derivative

X : $O-CH_2-CH_2$; CH_2-CH_2

Scheme 2.

Table 3. Rate constants and activation parameters for increase in carboxyl group content on thermal degradation of poly(alkylene terephthalates) [catalyst: Ti(butylate)$_4$; in PET: Sb- or Zn-derivative].

	$k \cdot 10^6$, s^{-1}				E_A kJ mol^{-1}	ln A (s^{-1})	$\Delta S^{\neq}$ J mol^{-1}K^{-1}
	250°C	260°C	265°C	280°C			
PBT open system	0.21	0.38	0.54	1.52	162	21.87	−76.9
PBT closed system	0.63		1.72	5.09	168	24.36	−55.2
PBT /21/	0.40			3.45	(171)	(24.73)	(−52)
PBT /22/	0.44	0.59		2.24	(142)	(17.87)	(−26)
PET 1				0.42	187	25.84	−41.7
PET 3				0.65	153	19.1	−100
PHMT /23/		0.60		1.69			
PDMT /23/		0.64		1.00			

scission of a proton and the formation of vinyl group (b). At longer distances between the activated carbonyl group and the ester bond to be broken, as in DB or BB, the interaction of a metal ion with the carbonyl group will not accelerate the reaction (c).

In thermal decomposition of PBT the primary step is likewise a chain scission resulting in a carboxyl end-group and an unsaturated ester end-group [19], but (as we could prove) without acceleration by metal compounds. In the resulting reactions 1-hydroxybutene-3, butadiene and large amounts tetrahydrofuran are formed [19,20]. From the results of the model reactions, a considerable lower thermal stability of PBT compared with PET is to be expected. The data presented in Table 3 verify these results. The rate constants and the activation parameters of PBT degradation correspond very well with the results of the model reaction, thus supporting the conclusions from the discussion on the reaction mechanism. Results with poly(hexamethylene terephthalate) and poly(decamethylene terephthalate) [23] are in the same direction. No particular effect of the catalyst on the rate of thermolysis was observed with polyesters of long-chain diols.

Now, with the knowledge of this thermal behavior of polyalkylene terephthalates, what can we do to retard degradation and to obtain polyesters which exhibit good thermal stability?

The results described indicate that initiation of degradation at the chain ends is unlikely. Hence it is not possible to improve the thermal stability by masking hydroxyl end-groups. On the contrary, in order to preserve the molecular weight over prolonged times of thermal load the polycondensation has to be performed in such a way that end-groups corresponding to a given degree of polymerisation remain as hydroxyl groups as long as possible. It is only in that way that the PET chain is regenerated after thermal scission. Therefore the catalytic effect of metal compounds on the degradation process has to be reduced as far as possible. Sb or Ge compounds have only a slight disadvantageous influence on thermal stability but they do not catalyse the ester interchange of DMT with ethylene glycol. The metal derivatives particularly suitable for this catalytic action (Zn, Co or Mn compounds) are also particularly active in accelerating chain scission, or cause other undesirable effects. These difficulties have been overcome by using catalyst/inhibitor combinations [1,7]. After performing the transesterification of DMT in the presence of a Co or Mn catalyst, a two- or threefold excess of a phosphorus compound such as an ester of phosphorus acid is added. In this way the metal derivatives are converted to insoluble and catalytically inactive phosphates (see sample 2 in Tables 1 and 2). Then the polycondensation can be performed using the catalytic activity of Sb, Ge or Ti compounds which is not affected by the phosphorus compound. Thus phosphorus compounds do not actually alter the thermal stability of PET itself, but prevent the additional acceleration of thermal degradation caused by metal catalysts or other metal traces.

In thermal degradation of PBT, contrary to PET, no accelerating influence of metal compounds is detectable. Thus we can conclude that in thermal degradation of PBT and polyesters with long chain diols, the reaction temperature plays the predominant role and it is impossible to increase the stability of these polyesters by addition of metal blocking agents. To shorten the thermal loadtime during the polycondensation of PBT, the catalyst concentration should be high and the reaction temperature as low as possible. There are, however, indications of a decrease in catalytic activity of Ti(butylate)$_4$ with increasing concentration caused by an autoassociation of the catalyst [24].

This temperature effect has to be taken into account, e.g., in the production of glass fiber-reinforced PBT. In a compounding process, energy is conducted into the product in the form of internal shear generated by the mixer and thermal conditions of the polymer matrix may become critical. By addition of unmolten polymer into the mixing stage of a kneader the product end-temperature and consequently thermal degradation of a temperature sensitive matrix material like PBT can be reduced and

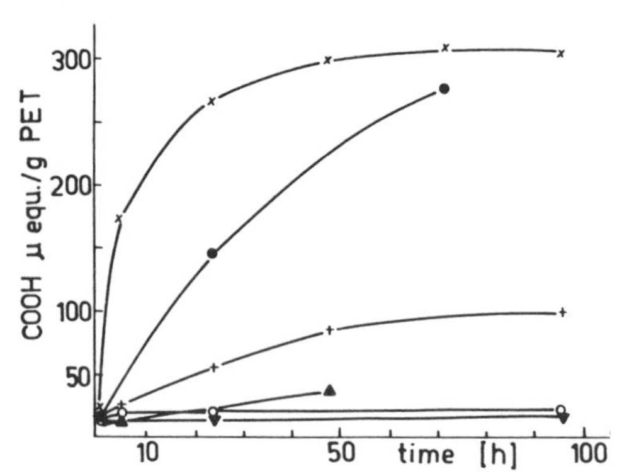

Figure 1. Increase in carboxyl groups of PET at 160°C in air in dependence on

Catalyst	and	DEG wt. %
+ Ge-derivative		1.0
× Ge-derivative		5.6
○ Sb-derivative		0.4
▼ Sb-derivative		1.4
▲ Sb-derivative		2.2
● Sb-derivative		4.0

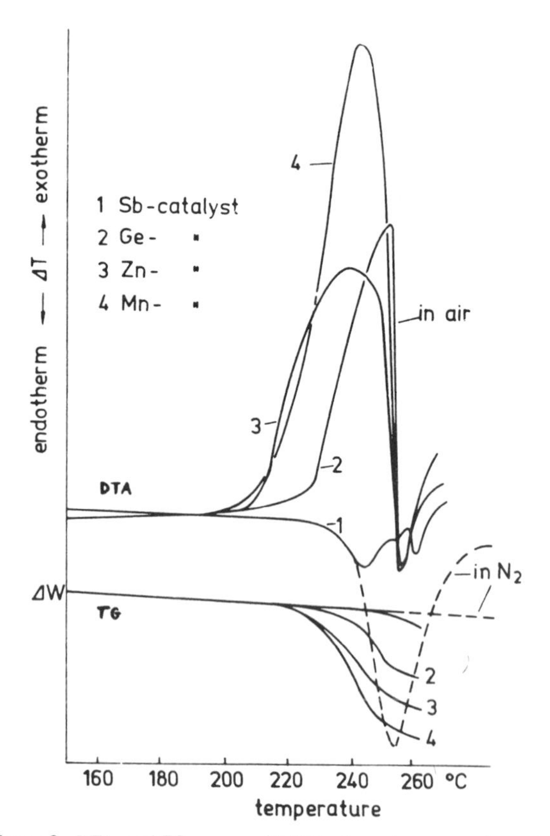

Figure 2. DTA- and TG-curves of PET in dependence on polycondensation catalysts (q: 4 K/min).

thereby the temperature-residence time relationship can be optimized [25].

Thermooxidative Degradation

Obviously, thermooxidative degradation proceeds by a rather complex mechanism [11,26–29]. It can be concluded that at least in the early stages (as in thermal degradation), the formation of carboxyl end-groups is a predominating reaction. Secondary reactions of radicals can result in the formation of network structures [26,28, 29] and by thereby the formation of gel-like particles in the polyester, which may yield mechanical inhomogeneity in the preparation of films or fibers [30].

Contrary to the thermal decomposition of PET, thermooxidative degradation can occur in the solid state to a considerable extent. We can prove that thermooxidative stability of PET also depends largely on the nature of transesterification catalysts, as well as upon the chemical structure of the polymer chain—especially on the content of ether linkages (DEG).

Figure 1 shows the increase in carboxyl groups by heating PET at 160°C in air. Sb-containing samples with low content of DEG are over a prolonged time, as stable as in N_2. However, with increasing content of DEG thermal oxidative degradation also increases (quite in contrast to thermal degradation). Ge-containing PET samples obviously are much more sensitive to thermooxidative degradation.

In thermoanalytical investigations, differences in the thermooxidative stability of PET are observed from the onset and the extent of an exothermic DTA-peak prior to melting of the polyester as well as by differential mass losses (Figure 2).

From the difference of the DTA exothermic and endothermic peaks obtained by heating a PET sample in air and nitrogen, a relative enthalpy of oxidation can be calculated. These values and the onset temperature of the exotherm are very suitable for characterising the thermooxidative stability of PET [30]. Table 4 summarises such values.

Sample 1 is far more sensitive to oxidation than sample 2. The difference results only from the method of their preparation. While sample 1 was prepared by esterification of terephthalic acid, resulting in a rather high amount of DEG; sample 2 was obtained by transesterification of DMT, using a catalyst/inhibitor combination. Other samples show the different oxidation promoting effect of the individual metal compounds if these are not (samples 3 and 8) or not sufficiently (sample 5) blocked by a catalyst inhibitor—or if this means of deactivation of metal traces is not possible (Ge, Ti compounds). Samples 9, 10, 11 and 12 demonstrate how much the modification of the chemical structure of PET may influence the thermooxidative stability by introducing small quantities of co-components into the polymer backbone chain.

Table 4. Onset temperature of oxidation T_{Ox}, enthalpy of oxidation ΔH_{Ox} and mass loss Δm of PET by nonisothermal heating in air up to 250°C in dependence on catalysts and additives (heating rate q:4 K/min).

PET	Catalyst/Inhibitor 10^{-4} wt.% in PET	DEG wt.%	T_{Ox} °C	ΔH_{Ox} kJ mol^{-1}	Δm %
1	139 Sb	2.54	204	24.4	0.48
2	41 Mn/ 216 Sb/ 42 P	0.6	248	3.9	0.10
3	92 Zn	1.3	201	53.0	0.74
5	100 Mn/ 248 Sb/ 10 P	1.1	198	56.2	1.04
6	70 Mn/ 76 Ge/ 79 P	1.1	216	34.2	0.49
7	68 Ti	0.9	214	33.6	0.68
8	86 Co	1.9	199	56.3	1.42
9	82 Mn/ 170 Sb/ 31 P	2.75	191	62.3	1.43
10	143 Sb/ 1.7 mole % Na-sulfo-isophthalate	7.45	185	118*	1.61*
11	149 Sb/ 1.7 mole % Na-sulfo-isophthalate	3.66	214	21.1	0.18
12	78 Mn/ 161 Sb/ 71 P 5 mol % poly(ethylene oxide) 4000		103	984	30

*Up to 240°C.

Segmented polyether-ester copolymers like sample 12 display many of the physical characteristics of cured elastomers, such as impact resistance and low-temperature flexibility and yet they can be processed using conventional thermoplastic equipment and procedures.

Although polyether-ester and polyether-amide copolymers are of increasing commercial importance, however, the thermoanalytical data indicate extremely oxidative sensibility of these block copolymers. By addition of antioxidants, thermooxidative degradation of such compounds can be delayed to a large extent [31].

The segmented copolyesters of poly(alkyleneoxides) and PBT exhibit a thermooxidative behavior similar to the PET copolyesters. They must also be stabilised by antioxidants. The commercial thermo-elastoplastic copolyetheresters *Hytrel* and *Elitel* have good thermal stability in inert atmosphere. The thermoanalytical curves (Figure 3) indicate however that they nevertheless have to be handled with care at higher temperatures in the presence of air.

Many patents propose the additon of antioxidants to PET, but with the homopolymer it is not absolutely necessary and it is seldom done in technical practice. Some problems are connected with the use of antioxidants. Their efficiency in PET depends strongly upon their chemical structure as well as on the moment of addition to the polyester. A temperature of about 280°C and a high vacuum during polycondensation can cause loss of antioxidants through volatilisation. Furthermore, the chemical structure of the antioxidant may be altered by reaction with the polyester or with reaction products of polycondensation and less effective or more volatile products can be formed.

Some examples of antioxidant activity are illustrated with Table 5. The commercial stabiliser Irganox 1222 can operate in two ways, as an antioxidant and as a metal blocking agent due to the phosphate part. Additionally, in this way the antioxidant would be fixed in the polyester. On account of the joint addition of triphenylphosphate, which reacted in competition with the metal this effect is not completely realised with sample B in contrast to samples C and H.

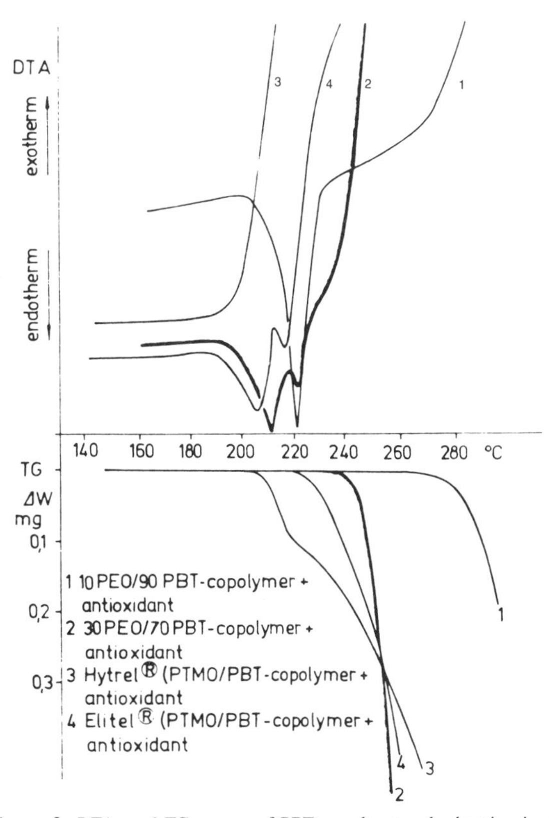

Figure 3. DTA- and TG-curves of PBT copolyesters by heating in air (q: 4 K/min).

Table 5. Onset temperature of oxidation T_{Ox}, enthalpy of oxidation ΔH_{Ox}, and mass loss of PET by thermooxidative degradation in dependence on catalysts and antioxidants (heating rate q: 4 K/min). Stability time S_t detected by isothermal heating in air. CDK: cyclohexanone diphenylamine condensate; DG: dodecyl-3,4,5-trihydroxybenzoate Irg. 1222: Irganox 1222; TPPat: triphenylphosphate.

PET	Catalyst/Inhibitor 10^{-4} mol mol^{-1} PET	Antioxidant wt.%	DEG %	T_{Ox} °C	ΔH_{Ox} kJ mol^{-1}	Δm %	S_t, min 220°C	S_t, min 240°C
A	3 K_2GeO_3		1.1	216	34.6	0.50		
A/m	3 K_2GeO_3	0.5 CDK	1.1	—	—	—		1,500
B	3 Mn ac_2; 3 K_2GeO_3 / 4 TPPat;	0.15 Irg. 1222 p.p.	1.2	232	16.3	0.17		
C	3 Mn ac_2; 3 K_2GeO_3	0.15 Irg. 1222 p.p.	1.0	249	—	0.1		
D	3 Mn ac_2; 3 K_2GeO_3 / 4 TPPat;	0.15 Irg. 1222 a.p.	1.1	—	—	—	760	78
E	3 Mn ac_2; 3 K_2GeO_3 / 4 TPPat;	0.2 DG a.p.	0.55	—	—	—	118	29
F	3 Mn ac_2; 3 K_2GeO_3 / 4 TPPat;		2.47	205	46.0	0.61		
F/m	3 Mn ac_2; 3 K_2GeO_3 / 4 TPPat;	0.2 DG	2.47	—	—	—		
G	3 Mn ac_2; 1.5 Sb_2O_3 / 2 TPPat;		2.75	191	62.3	1.43		
G/m	3 Mn ac_2; 1.5 Sb_2O_3 / 2 TPPat;	0.2 DG	2.75	—	—	—		
H	3 Mn ac_2; 1.5 Sb_2O_3	0.15 Irg. 1222 p.p.	0.41	248	3.9	0.16		

a.p.—antioxidant added after polycondensation.
p.p.—antioxidant added prior to polycondensation.
/m—antioxidant mixed with PET powder.

These examples prove thermoanalysis to be an excellent and convenient tool for the investigation of the thermooxidative stability of polyesters and the factors affecting it. Moreover, the results presented in Table 6 indicate that the exothermic DTA peak really reflects considerable degradation causing strong changes in specific parameters of the polymer. For this comparison thermoanalytical experiments were quenched at a certain temperature beyond T_{Ox}. η_{rel}, the carboxyl group content and ΔH_{Ox} evolved up to this time were estimated [12].

In the study of polymer decomposition the derivation of kinetic data using DTA and TGA has received increasing attention, as well as with much criticism, e.g. [32, 33]. Several theoretical approaches to the treatment of thermoanalytical data obtained by isothermal or non-isothermal measurements as well as various problems in the application of these methods to condensed phase polymer systems have been extensively discussed in the literature [32–34].

Table 6. ΔH_{Ox}, increase of COOH groups and change in η_{rel} of PET sample 5 in dependence on the temperature of interruption of the DTA measurement T_I and the particle size of the sample.

	Particle size, mm					
	0.12–0.25			0.25–0.50		
T_I °C	ΔH_{Ox} kJ mol^{-1}	COOH μeq g^{-1} PET	η_{rel}	ΔH_{Ox} kJ mol^{-1}	COOH μeq g^{-1} PET	η_{rel}
		38	1.390		38	1.390
213	3.0	123	1.274	2.6	94	1.292
221	8.9	230	1.199	7.1	163	1.250
230	22.8	436	1.134	13.5	244	1.202
245	41.5	619		29.1	462	

The study of the degradation kinetics of organic polymers like PET, pose various additional problems. As already emphasized such degradation reactions should be followed only to limited, rather low degrees of conversion. In thermogravimetric investigations with increasing temperature or increasing degree of conversion without any change in reaction mechanism, a change in the composition of degradation products may result and thus the basis of calculation would be altered.

All kinetic studies utilise the rate Equation (2) (α = degree of conversion)

$$\frac{d\alpha}{dt} = kf(\alpha) \tag{2}$$

Regarding the Arrhenius equation, and with $f(\alpha) = (1 - \alpha)^n$ in the case of a constant heating rate q results

$$\frac{d\alpha}{dT} = \frac{A}{q} e^{-E/RT}(1 - \alpha)^n \tag{3a}$$

or

$$\int_0^\alpha \frac{d\alpha}{(1 - \alpha)^n} = \frac{A}{q} \int_{T_0}^T e^{-E/RT} \tag{3b}$$

The relationship of Equation (3) is the basis of subse-

Table 7. Activation parameters and rate constants for thermooxidative degradation of PET calculated from nonisothermal TG measurements by the Zsako/Šatava method (q: 0.5 K/min; weight loss ≤ 1.5%). In brackets: calculated from DTA (Kissinger method).

PET	E kJ mol^{-1}	ln A (s^{-1})	$k \cdot 10^5$, s^{-1} 200°C	210°C	220°C	230°C	240°C
1	225	42.49		0.13	0.41	1.23	3.52
2	210	37.25				0.24	0.64
3	117	16.47	0.17	0.32	0.58	1.0	
5	153	26.30	0.34 (0.30)	0.75 (0.77)	1.63	3.43	
6	122	17.30	0.1	0.21 (0.11)	0.39 (0.42)	0.7	1.24
7	103	13.13	0.21	0.37	0.62	1.0	
PBT A	192	33.48	0.6	1.52			

quent analytical procedures, used to derive kinetic parameters from nonisothermal thermoanalytical data.

A differential method developed by Kissinger [35] involves the determination of the temperatures T_m of the maximum of the first derivative weight loss curve or of the DTA peak at different heating rates. We used the Kissinger method in a modified manner. The kinetic data on thermooxidative degradation of PET obtained agree very well with data evaluated from isothermal chemical investigations on the increase in carboxyl end-groups [12].

In the relationship (3b) the integral term on the right side can be resolved only by approximation methods (e.g., [36]). Šatava [37] proved that log $g(\alpha)$ vs. $1/T$ gives a straight line. From the slope of this line the overall activation energy E and A can be evaluated. In such a way it is possible to obtain from one nonisothermal T curve the kinetic parameters [$g(\alpha)$: conversion integral in Equation (3b)]. Figure 4 shows the Satava plot lg $g(\alpha)$ vs. $1/T$ of the results of nonisothermal thermogravimetric measurements and Table 7 presents the calculated kinetic data [38]. The different factors affecting the thermooxidative stability are clearly to be seen in the rate constants. For comparison some data evaluated from the DTA curve using the Kissinger method are included. They are in rather good agreement.

Photochemical degradation of PET has been extensively investigated (see, e.g. [39]). In order to examine whether there is an influence of residual metal traces similar to thermal degradation reactions we made some

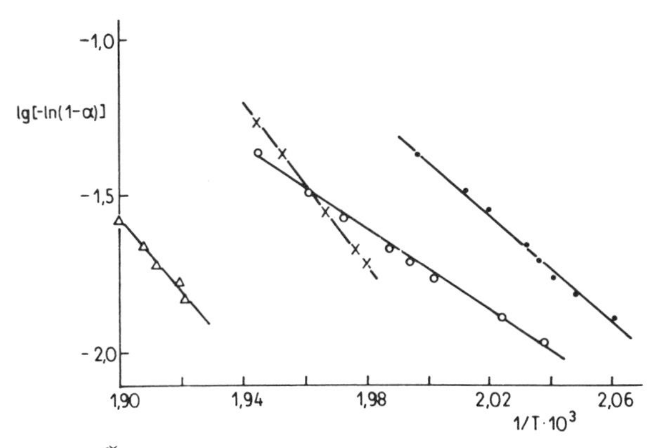

Figure 4. Šatava plot of weight loss by thermooxidative degradation of PET samples: × 1; △2; ○3; ●5 (1q: 0.5 K/min).

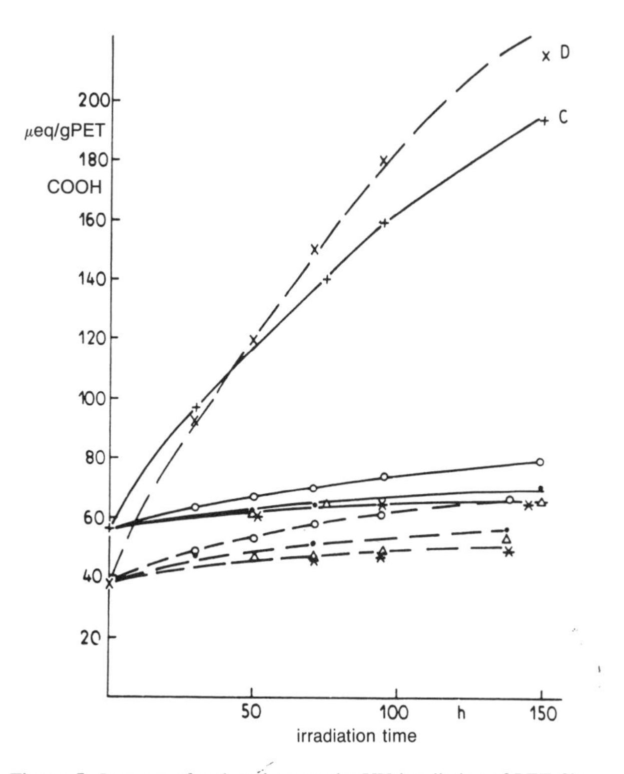

Figure 5. Increase of carboxyl groups by UV irradiation of PET film stacks (film layer 1 × +, 2○, 3●, 4△, 5∗). PET C film 23 µm thick; Co/Sb catalyst. PET D 20 µm; catalyst see sample 2.

preliminary experiments. Using stacks of 20 μm films we found only small differences between different PET samples without concordance to thermooxidative stability (T_{Ox}, ΔH_{Ox}), but observed a strong dependence on the film thickness (Figure 5).

CONCLUSIONS

The results demonstrate how much thermal stability and other properties of poly(alkyleneterephthalates) are affected by small quantities of impurities, additives or chain modifications. Taking these factors into account, improvements in stability can be attained. Therefore it is not sufficient to examine the degradation of the pure polymers; it must be studied in the presence of additives or known impurities.

It was illustrated that thermal analytical results reflect the real degradation reactions, and it is possible to get reasonable kinetic data and information on thermal stability from nonisothermal measurements.

REFERENCES

1. Zimmermann, H. in *Polyesterfasern, Chemie und Technologie*. 2nd ed. H. Ludewig, ed. Berlin:Akademie Verlag, Chapter 4 (1975).
2. Pohl, H. A. *Analyt. Chem.*, 26:1614 (1954).
3. Zimmermann, H., H. Hoyme and A. Tryonadt. *Faserforsch, Textiltechn.*, 21:33 (1970).
4. Hoyme, H., A. Seganova and H. Zimmermann. *Faserforsch. Textiltechn.*, 22:419 (1971).
5. Zimmermann, H. and D. Becker. *Faserforsch. Textiltechn.*, 24:479 (1973).
6. Zimmermann, H. and P. Lohmann. *Acta Polym.*, 31:686 (1980).
7. Zimmermann, H. *Faserforsch. Textiltechn.*, 13:481 (1962).
8. Ritchie, P. D. *Monograph No. 13*, Soc. Chem. Ind., London, p. 107 (1961).
9. Goodings, E. P. *Monograph No. 13*, Soc. Chem. Ind., London, p. 211 (1961).
10. Zimmermann, H. and E. Leibnitz. *Faserforsch. Textiltechn.*, 16:282 (1965).
11. Buxbaum, L. *Angew. Chem.*, 80:225 (1968).
12. Zimmermann, H. in *Developments in Polymer Degradation, Vol. 5*. N. Grassie, ed. London/New York:Appl. Sci. Publ., 79–119 (1984).
13. Halek, G. W. *J. Polym. Sci., Polym. Symp.*, 74:83 (1986).
14. Onasch, J. *Kunststoffe*, 76:481 (1986).
15. Kirshenbaum, G. S., W. T. Freed, M. W. Dong and J. T. Carrano. *Preprints of Org. Coat. and Pt. Chem., ACS National Meeting, Washington, DC, Sept. 1979.*
16. Kanchiku, Y. and N. Ohsuga. *Bunseki Kagaku*, 28:508 (1979).
17. Tomita, K. *Polymer*, 14:50 (1973).
18. Jabarin, S. A. and E. A. Lofgren. *Polym. Eng. Sci.*, 24: 1056 (1984).
19. Lum, R. M. *J. Polym. Sci. Chem. Ed.*, 17:203 (1979).
20. Möller, B., J. Blaesche, M. Mudrick, G. Rafler, H. Zimmermann and M. Stromeyer. *Acta Polym.*, 33:38 (1982).
21. Passalacqua, V., F. Pilati, V. Zamboni, B. Fortunato and P. Manaresi. *Polymer*, 17:1044 (1976).
22. Rafler, G. and J. Blaesche. *Acta Polym.*, 33:472 (1982).
23. Rafler, G., J. Blaesche, B. Möller and M. Stromeyer. *Acta Polym.*, 32:608 (1981).
24. Fortunato, B., P. Manaresi, A. Munari and F. Pilati. *Polym. Commun.*, 27, 29 (1986).
25. Stade, K. H. *Polym. Eng. Sci.*, 18:107 (1978).
26. Yoda, K., A. Tsuboi, M. Wada and R. Yamadera. *J. Appl. Polym. Sci.*, 14:2357 (1970).
27. Michailov, N. V., G. M. Terechova, L. G. Tokareva and N. G. Karkova. *Chim. Volokna*, 4:15 (1970).
28. Nealy, D. L. and L. J. Adams. *J. Polym. Sci.*, A1(9):2063 (1971).
29. Spanninger, P. A. *J. Polym. Sci. (Polym. Chem. Ed.)*, 12: 709 (1974).
30. Zimmermann, H., D. Becker and E. Schaaf. *J. Appl. Polym. Sci., Appl. Polym. Symp.*, 35:183 (1979).
31. Zimmermann, H. and K. Dietrich. *Acta Polym.*, 30:199 (1979).
32. Still, R. H. in *Developments in Polymer Degradation, Vol. 1*. N. Grassie, ed. London:Applied Science Publishers Ltd., p. 1 (1977).
33. Schneider, H. A. in *Degradation and Stabilization of Polymers, Vol. 1*. H. H. G. Jellinek, ed. Amsterdam-Oxford-New York:Elsevier, 506–553 (1983).
34. Behnisch, J., E. Schaaf and H. Zimmermann. *J. Thermal Anal.*, 13:117, 129 (1978); 15:285 (1979).
35. Kissinger, H. E. *Anal. Chem.*, 29:1702 (1957).
36. Zsako, J. *J. Phys. Chem.*, 72:2406 (1968).
37. Šatava, V. *J. Thermal Anal.*, 5:217 (1973).
38. Zimmermann, H. and J. Behnisch. (To be published.)
39. Day, M. and D. M. Wiles. *J. Appl. Polym. Sci.*, 16:175, 191, 203 (1972). S. Krishnan, S. B. Mitra, P. M. Russell and G. Benz. *ACS Symposium Series No. 287*, 389 (1985).

L. B. CARETTE[1]
M. GAY[2]

Optimization of Stabilizing System for Polyolefins

ABSTRACT

Statistical method for stabilizing efficiency evaluation of a diphenolic antioxidant bis [2,2′ methylene bis(4 methyl 6 t.butyl phenol)]terephthalate explains the differences which are observed if two phenolic antioxydants are used under the same experimental conditions including the same concentrations, and the same process conditions. Trials carried out on polypropylene stabilized with mixtures of antioxydants, phosphites, with thioesters or with no stabilizer have shown that some phenomena cannot be deduced from theoretical aspects but implicate physical phenomena which must be connected with the molecular structure independently of the chemical activity.

Likewise the synergistic effect on oven aging of sulfur compounds/phenolic antioxydants mixtures can be explained at the same time with the theoretical chemical mechanism and the physical improvement of the antioxydant dispersibility.

KEY WORDS

Statistical method, phenolic antioxydant, polypropylene, mixture optimization, oven aging, process stabilization, dispersibility, compatibility.

1. INTRODUCTION

The mechanisms of antioxydant action are widely described by numerous authors. These theoretical reflections are limited from a practical point of view and it is necessary to perform trials with mixtures of stabilizers. Usually the practical tests concern the process stability, and the thermal stability. The results are commonly given in tables where some groups of formulations are collected.

Efficiency comparison of different stabilizers is made at the same levels of concentrations—sometimes the conclusions are encouraging and other times disappointing.

As a method of evaluation of the phenolic compound: bis [2,2′ methylene bis(4 methyl 6 t.butylphenol)] terephthalate (AO I) we have decided to perform experiments according to a statistical method which permits coverage of the useful range of concentrations of the chemical additives used. The main advantage of this method is to reduce the number of trials for the largest evaluation.

The results are given in polynomial form to show interactions between additives and optimum combinations of concentrations.

The torque rheometer was chosen to evaluate the processability.

Concerning the oven life of the polypropylene, a general view of a lot of trials performed with the AO I leads to other experiments such as DSC to demonstrate the part of thioester increasing the fusibility of the AO, thus enhancing the synergistic effect.

2. ADDITIVES USED IN EXPERIMENTS

In the following experiments we have used the AO I antioxydant described above: the tris (2,4 dit.butylphenyl) phosphite, the distearoylthiodipropionate DSTDP, and as standard reference the pentaerythrityl-tetrakis (3(3,5 di t.bu. 4 hydroxyphenyl) propionate)—AO II.

AO I
OH O—CO—⟨⟩—CO—O OH

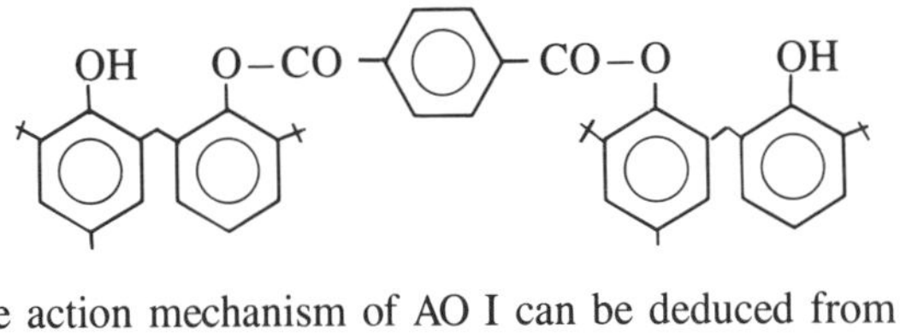

The action mechanism of AO I can be deduced from what we know about the mechanism of another diphenolic compound: 2,2 methylene bis (4 methyl 2 tertiarybutyl phenol) AO III

AO III
OH OH

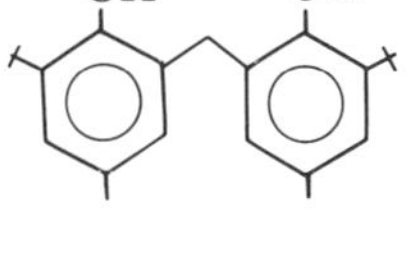

[1]Rhone Poulenc—Recherches, Centre de Recherches D'Aubervilliers, 93308 Aubervilliers, France.
[2]Rhone Poulenc—Recherches, Centre de Recherches des Carrieres, 69192 Saint Fons, France.

The structure evolution of AO III and the compounds which are formed during the stabilizing action are known and described in following Schemes 1, 2, 3.

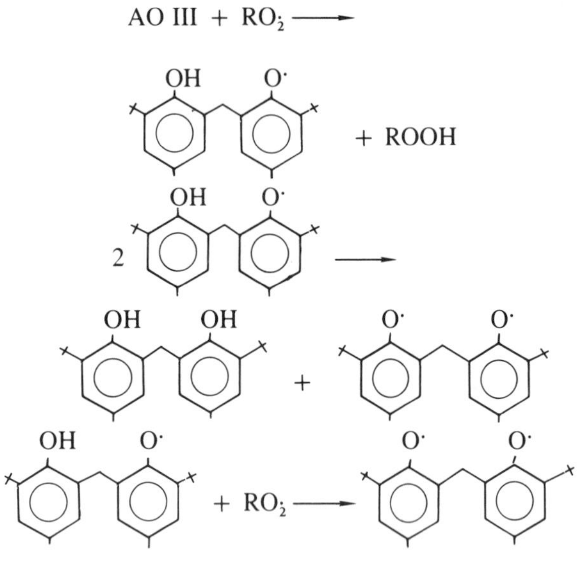

AO III + $RO_2^{\cdot}$ →

OH O· + ROOH

2 OH O· →

OH OH + O· O·

OH O· + $RO_2^{\cdot}$ → O· O·

I II

Scheme 1.

But the diphenoxy radical could give a quinone-methyde structure VII which is colored.

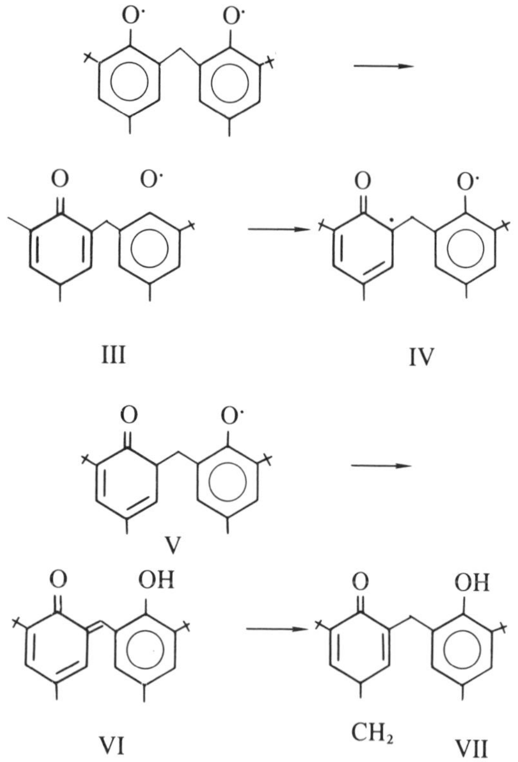

O· O· →

O O· → O O·

III IV

O O· →

V

O OH → O OH

VI CH_2 VII

Scheme 2.

From AO III it is possible to obtain phenolic dimers or trimers VIII which are colourless according to reactions in Scheme 3 but could give stilbene quinone IX structures much more coloured than the quinone methyde (Scheme 4).

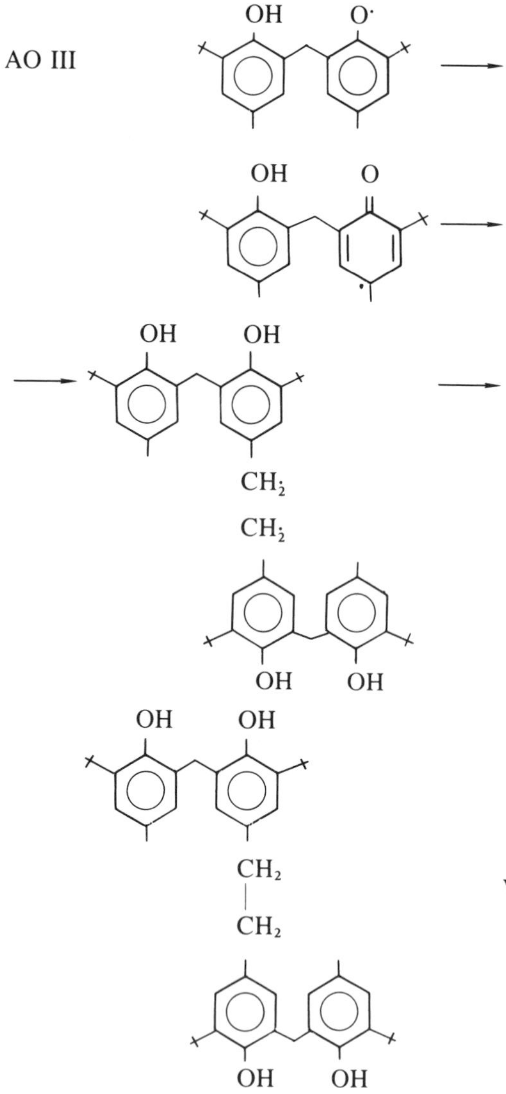

AO III OH O· →

OH O →

→ OH OH →

CH_2

CH_2

OH OH

OH OH

CH_2

CH_2 VIII

OH OH

Scheme 3.

OH OH

CH_2

CH_2

OH OH

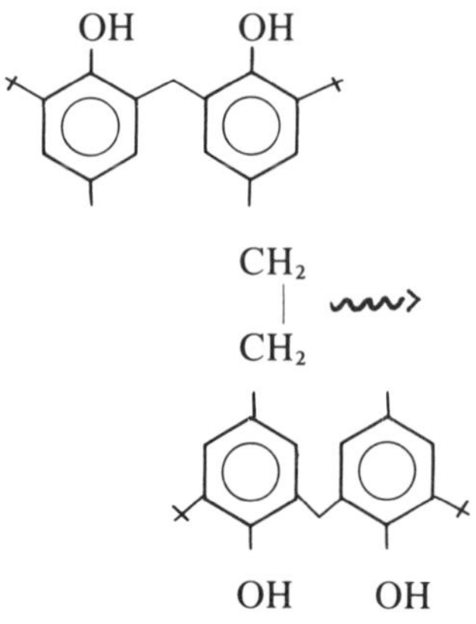

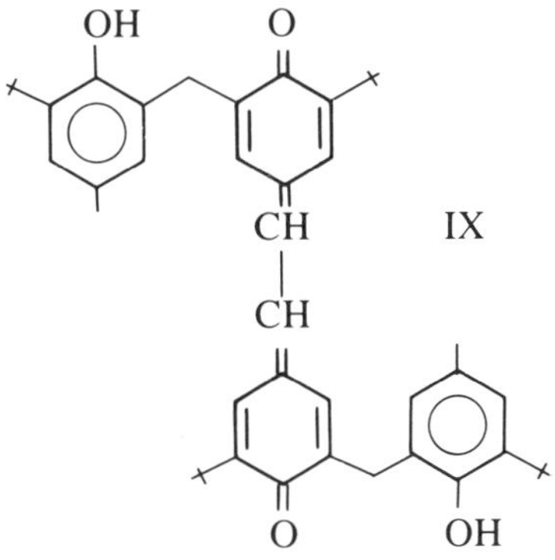

IX

Scheme 4.

The action mechanism of AO I is a classical one (Scheme 5) and like AO III give dimer structures X.

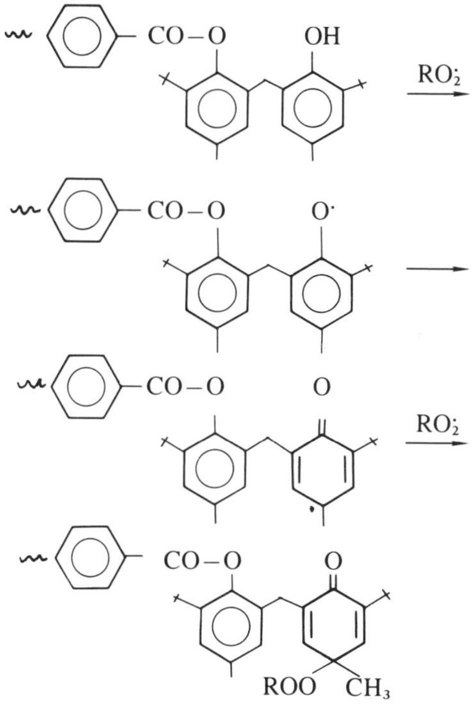

Scheme 5.

But as a consequence of the distance between the active phenolic sites, the probability of the reaction leading from V to VI is small. The chance to obtain coloured products is reduced.

On the other hand the phenoxy radicals which are formed on the dimer structure could lead to an internal transposition regenerating the phenolic groups (Scheme 6).

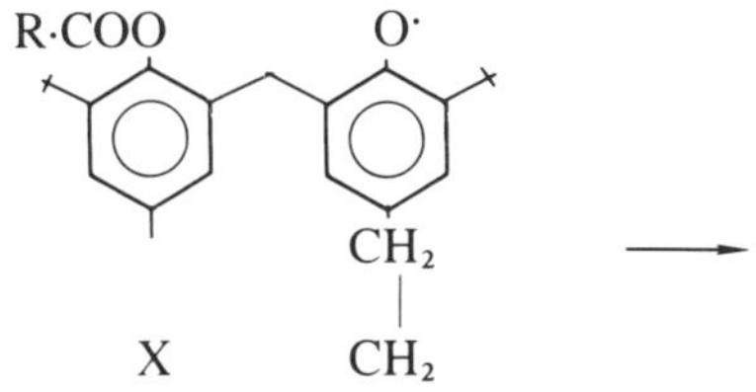

X

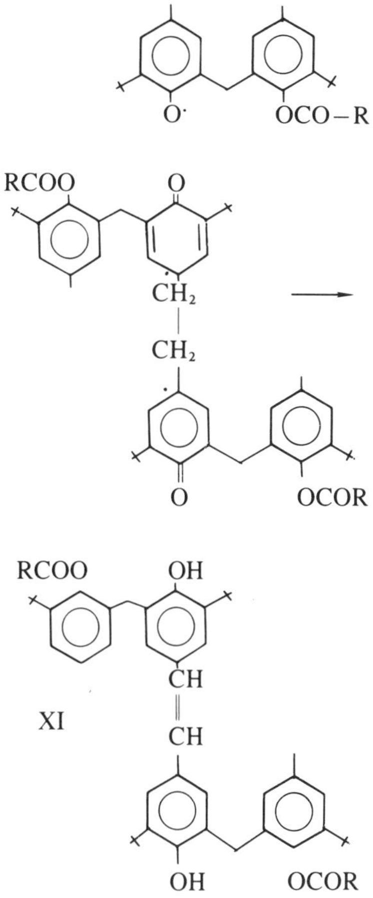

XI

Scheme 6.

But these assumptions do not allow us to predict the efficiency of AO I in a polymer matrix.

3. EXPERIMENTAL METHOD

3.1. Study of Process Stability

Process stability tests have been carried out using a laboratory mixer with the ability to measure the mixing torque (Brabender Plastograph). The mechanical energy given to the polymer for uniform time (20 minutes) has been chosen as the stabilization criterion. This criterion is convenient because polypropylene begins to degrade immediately under thermomechanical stress.

It would not have been desirable to chose as a criterion torque values at given times (5, 10, or 20 minutes) because of relatively high frequency irregularities in the measure of the torque. Calculating the integral of the torque versus time plot makes these irregularities ineffective. This integral correctly corresponds to the chosen criterion, i.e., the mechanical energy. The stabilization criterion is defined hereafter:

$$A = \sum (\text{Torque} \cdot \text{Time})$$

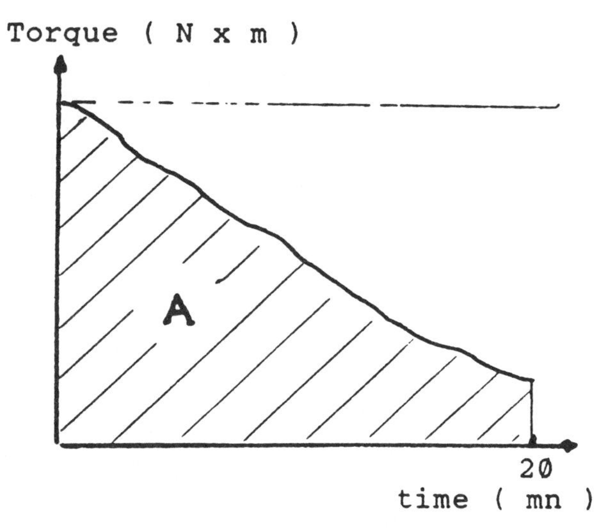

Figure 1. Process conditions: rotor speed—60 rpm; temperature—220°C.

It is interesting to notice that the torque values which have been taken into account to calculate the integral are corrected values, using an Arrhenius-type law. In fact, the temperature of the product is not constant during the test, especially at the beginning where it is lower. So it is important to recalculate the value of the torque as if the temperature were constant during the whole length of the test.

The use of a laboratory mixer for the evaluation of polyolefin stabilization, a method already described by some authors, is realized by the simultaneous use of a computer making real-time corrections and check-up of the test.

3.1.1. STASTICAL METHOD—CHOOSING FORMULATIONS

Different compositions using a phenolic antioxydant, a phosphite and a thioester at concentrations as described in Table 1 are tested.

Formulations are settled within a central composite programme built as shown hereafter in Table 2.

Some formulas are obviously useless from a practical point of view. But it is necessary to examine them so that analysing results leads to a mathematical representation which will be significantly expressed as a polynomial function.

Table 1.

	$+\alpha$	$+1$	0	-1	$-\alpha$
Antioxidant	0	0.009	0.05	0.091	0.1
Phosphite	0	0.02	0.1	0.18	0.2
Thioester	0	0.02	0.1	0.18	0.2

Table 2.

Test #	Antioxydant $X1$	Phosphite $X2$	Thioester $X3$
1	$+1$	$+1$	$+1$
2	$+1$	$+1$	-1
3	$+1$	-1	$+1$
4	$+1$	-1	-1
5	-1	$+1$	$+1$
6	-1	$+1$	-1
7	-1	-1	$+1$
8	-1	-1	-1
9	$+\alpha$	0	0
10	$-\alpha$	0	0
11	0	$+\alpha$	0
12	0	$-\alpha$	0
13	0	0	$+\alpha$
14	0	0	$-\alpha$
15	0	0	0

$\alpha = 1.216$.

3.1.2. RESULTS

Results may be expressed as following:

$$A = C_0 + \sum (C_i \cdot X_i) + \sum (C_{ii} \cdot X_i \cdot X_i)$$
$$+ \sum (C_{ij} \cdot X_i \cdot X_j) + \sum (C_{ijk} \cdot X_i \cdot X_j \cdot X_k)$$

where

X_1 = AO I concentration
X_2 = phosphite concentration
X_3 = thioester concentration

and displayed in three-dimensional charts.

Figures 2, 3 and 4 show energy given to the polypropylene according to the respective concentrations of antioxydant and phosphite for three particular values of the concentration of thioester (0%; 0.1%; 0.2%). Stabilization is all the more efficient since changes in viscosity have been minimized during the test and therefore the value of the energy is high.

This kind of figure allows us to make some observations:

1. Difference between the charts for the different concentrations of thioester is slight.
2. It is possible to determine concentrations of antioxydant and phosphite giving the best efficiency.

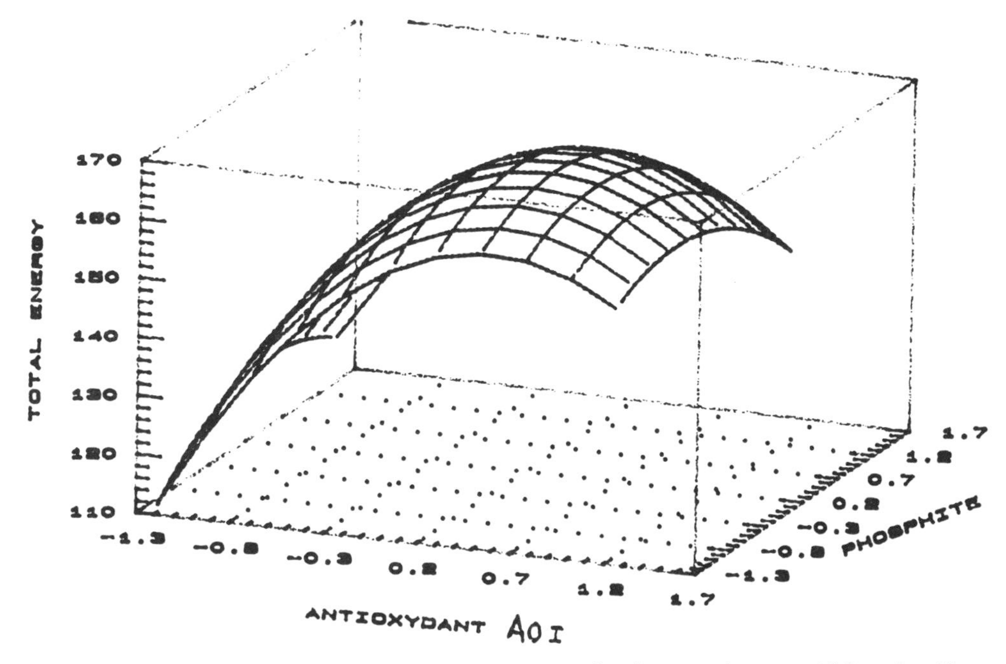

Figure 2. Efficiency of antioxydants mixtures expressed as energy given in torque rheometer. AO I + phosphite + 0% DSTDP.

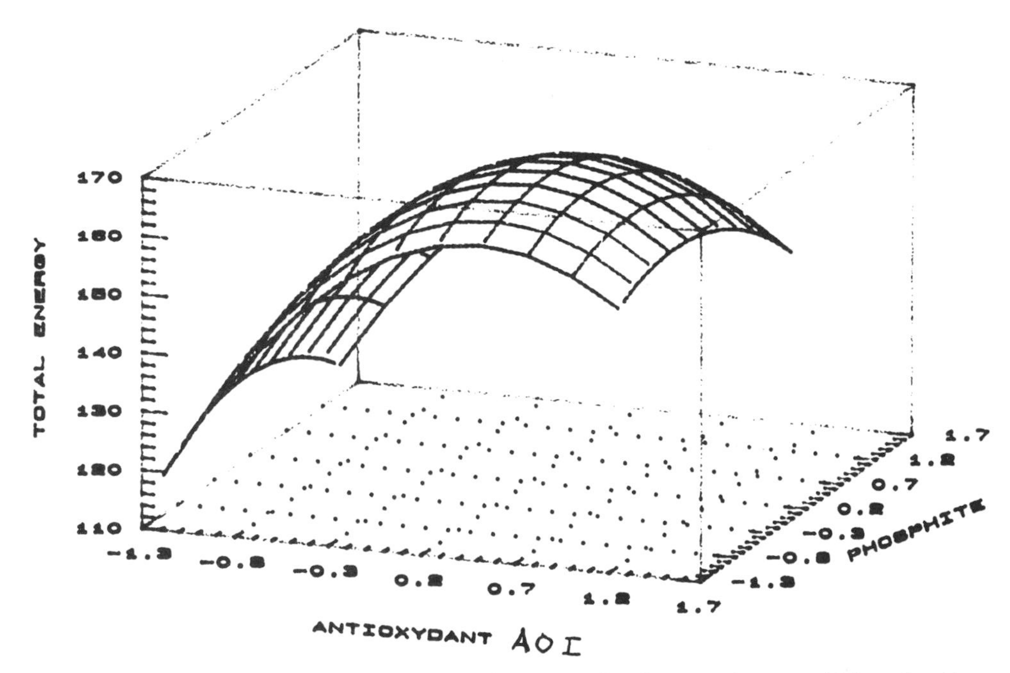

Figure 3. Efficiency of antioxydants mixtures expressed as energy given in torque rheometer. AO I + phosphite + 0.1% DSTDP.

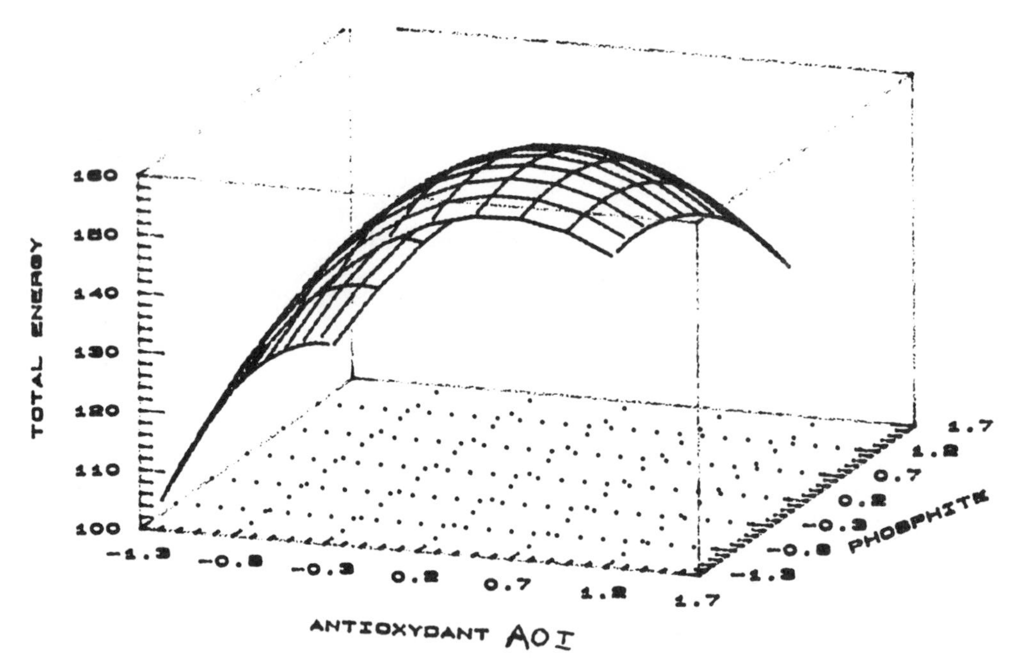

Figure 4. Efficiency of antioxydants mixtures expressed as energy given in torque rheometer. AO I + phosphite + 0.2% DSTDP.

3. Presence of thioester moves the optimum concentration of the phosphite. The higher the concentration of thioester is, the lower this optimum is. The concentration optimum of antioxydant is not significantly affected.
4. Comparison between antioxydants AO I and AO II is characteristic. Figure 5 shows the calculated surface for antioxydant AO II for a mean value of 0.1% of DSTDP. It is noticeable that the increase of energy is more continuous with a (probable) stagnation at higher values of antioxydant concentration, and that there is also an optimum for the concentration of phosphite.

We will make the following hypothesis: the de-

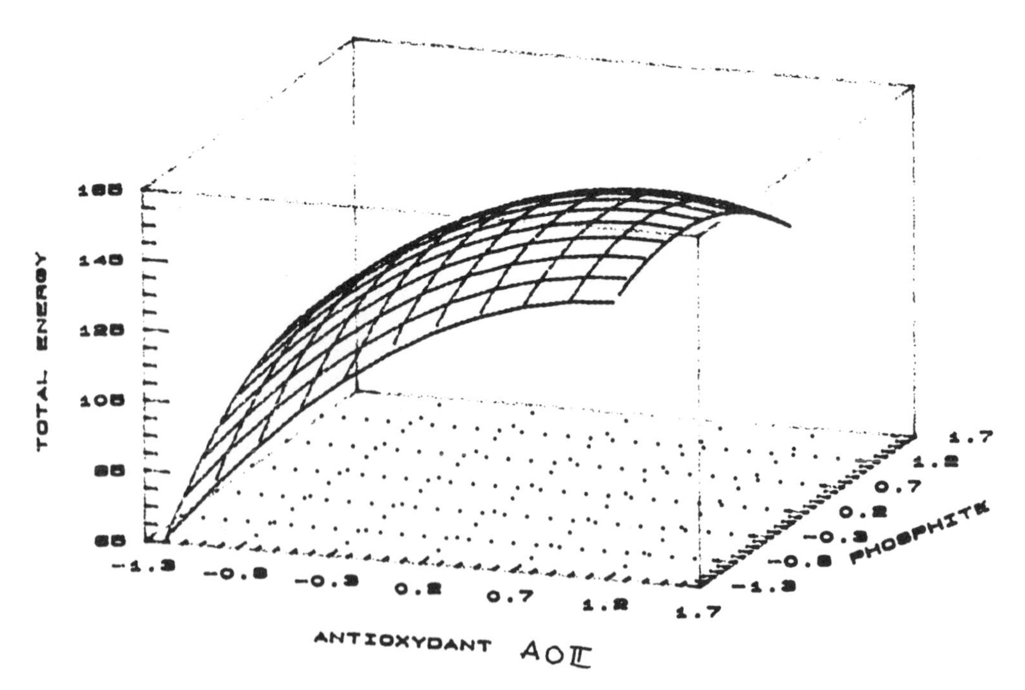

Figure 5. Efficiency of antioxydants mixtures expressed as energy given in torque rheometer. AO II + phosphite + 0.1% DSTDP.

Table 3.

System	Antioxidant AO I	Antioxidant AO II	Phosphite	Calculated Integral
First	0.1		0.2	150
		0.1	0.2	189
Second	0.05		0.1	190
		0.05	0.1	155

crease of efficiency for antioxydant AO I could be explained by two effects becoming competitive as its concentration is increasing:

- Efficiency is increasing in a normal way with the concentration of antioxydant.
- The probability to yield dimers is higher and higher. But the antioxydant is more difficult to disperse and its efficiency will not develop as well (molecular weight of more than 1800).

This second item will be discussed again in a later step about thermal aging.

5. It is difficult to compare products in strictly identical formulations. For example, if certain values of concentrations are chosen, values of the maximal energy will be obtained—and if other values of concentrations are chosen, values of energy will respectively be different.

For example, in Table 3 we could conclude that AO I is not so efficient as AO II with the first system of stabilizers.

On the contrary with the second system we obtain the opposite conclusion.

4. STUDY OF THERMAL STABILITY IN OVEN TEST

All the results described in this section have been obtained in different laboratories with polypropylene on 1.25 mm thick samples in ventilated oven tests at 150°C. If the aging conditions are similar the process procedures to make samples were different (strip extrusion, compression-molding, injection-molding).

The stabilizing system is made of AO I–DSTDP.

In Figure 6 are plotted the oven life (oven life is given by the embrittlement time) versus the DSTDP concentration at different levels of AO I concentration.

Figure 6 shows the upper limits which have been obtained but in fact at each AO I concentration, the dispersion of the results could be plotted as in Figure 7.

The survey of these charts shows:

- The synergism between AO I and DSTDP is very important although efficiency of AO I alone is very low.
- Some results are below the upper limit.

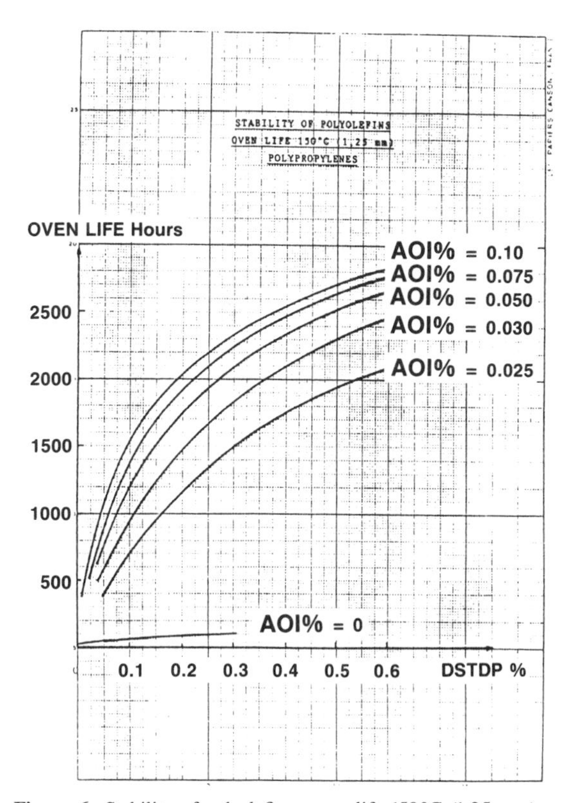

Figure 6. Stability of polyolefins—oven life 150°C (1.25 mm)—polypropylenes.

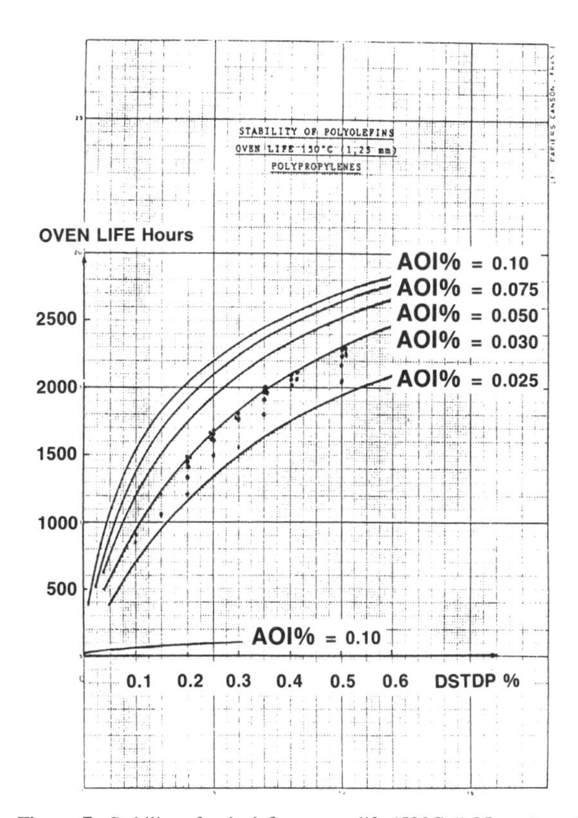

Figure 7. Stability of polyolefins—oven life 150°C (1.25 mm)—polypropylenes.

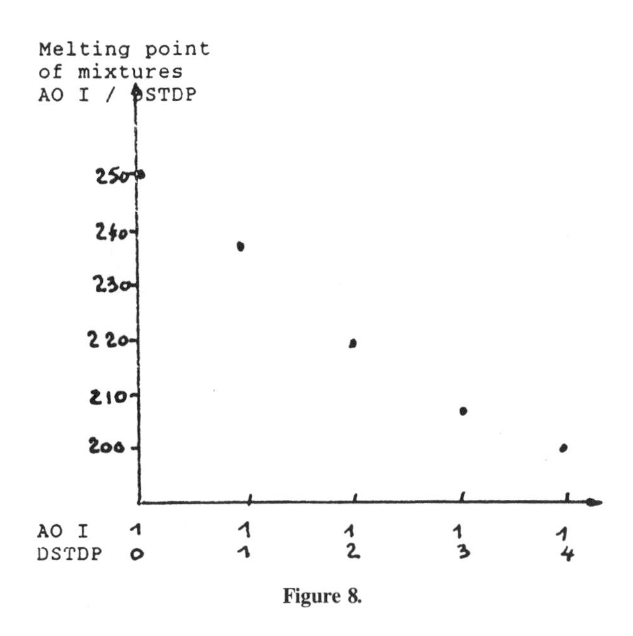

Figure 8.

In order to explain the strong synergism and the irregularities, we think that the more or less effective dispersion of the additives determines the quality of the results.

In presence of DSTDP it could be assumed at the melting point of AO I drops.

This fact is confirmed by measurements of melting points of mixtures made with different ratio AO I/DSTDP (Figure 8).

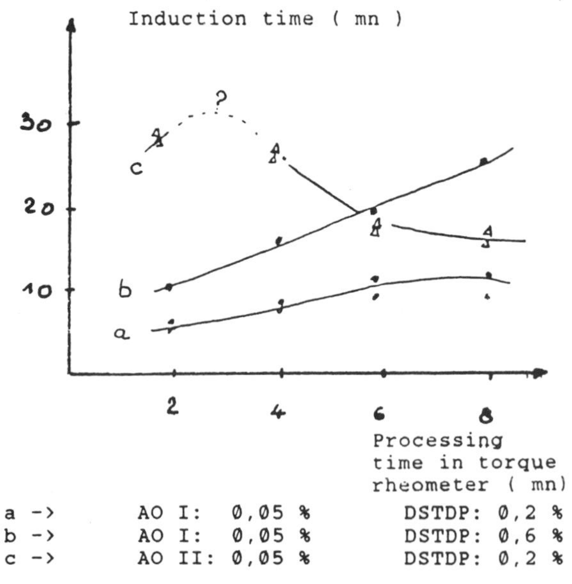

Figure 9. Oxidation induction time (DSC) versus processing time.

The importance of mixing conditions—as well as the role that the thioester could play during mixing on the dispersion and therefore on efficiency—may be demonstrated in measuring oxidation induction time with DSC on polypropylenes mixed for different times.

A Brabender plastograph was used to prepare the samples at 220°C, 60 rpm, for 2, 4, 6, 8, and 10 minutes.

Figure 9 shows the results obtained with the following formulations:

a)	AO I: 0.05%	DSTDP: 0.2%
b)	AO II: 0.05%	DSTDP: 0.6%
c)	AO II: 0.05%	DSTDP: 0.2%

We note that if the oxidation induction time decreases regularly from the beginning with AO II, it first increases with AO I to go through a maximum.

Increasing the concentration of DSTDP gives us an interesting result: the induction time has been dramatically improved and the maximum has been shifted, although it is admitted that this method is not able to detect the presence of thioester, but only of primary antioxydant.

These tests confirm the importance of physical parameters which should be taken into account when evaluating efficiencies of the stabilizers or perfecting a new one.

CONCLUSION

According to the performed tests, it has been shown that it is possible to obtain good levels of stability in polypropylene with the phenolic antioxydant AO I.

However, results could sometimes appear contradictory if tests are not numerous enough or not appropriately planned.

Only by examining a large number of formulations or by using a fitted statistical method is one able to account for the practical efficiency of a stabilizing system.

REFERENCES

1. Pospisil, J. *Die Angewandte Makromolecular Chemie*, 28: 13/29 (1973).
2. Pospisil, J. *Advances in Polymer Science*, 36:74/76 (1980).
3. Goodrich, J. E. and R. J. Porter. *Polymer Engineering and Science* (January, 1977).
4. Marshall, George and J. M. Turnipseed. *Polymer Eng. and Science*, 13(6) (November, 1973).